Bridge Resource Management for Small Ships

The salvage operation in full swing after the *Herald of Free Enterprise* rolled on its side and sank. (Scaldis)

Bridge Resource Management for Small Ships

The Watchkeeper's Manual for Limited-Tonnage Vessels

Captain Daniel S. Parrott

International Marine/McGraw-Hill

Camden, Maine • New York • Chicago • San Francisco • Lisbon • London • Madrid • Mexico City • Milan • New Delhi • San Juan • Seoul • Singapore • Sydney • Toronto

The McGraw·Hill Companies

Copyright © 2011 by Daniel S. Parrott. All rights reserved. Printed in the United States of America. Except as permitted under the United States Copyright Act of 1976, no part of this publication may be reproduced or distributed in any form or by any means, or stored in a database or retrieval system, without the prior written permission of the publisher. The name "International Marine" and the International Marine logo are trademarks of The McGraw-Hill Companies.

1 2 3 4 5 6 7 8 9 10 11 12 13 14 15   QDB/QDB   1 9 8 7 6 5 4 3 2 1
ISBN 978-0-07-1550079
MHID 0-07-155007-0
EBOOK ISBN 0-07-155008-9

Library of Congress Cataloging-in-Publication Data is available from the Library of Congress.

This publication is designed to provide accurate and authoritative information in regard to the subject matter covered. It is sold with the understanding that neither the author nor the publisher is engaged in rendering legal, accounting, securities trading, or other professional services. If legal advice or other expert assistance is required, the services of a competent professional person should be sought.
  —*From a Declaration of Principles Jointly Adopted by a Committee of the American Bar Association and a Committee of Publishers and Associations*

International Marine/McGraw-Hill books are available at special quantity discounts to use as premiums and sales promotions or for use in training programs. To contact a representative, please e-mail us at bulksales@mcgraw-hill.com.

This book is printed on acid-free paper.

Note: The author and publisher shall not be liable to any user of the book for any loss or injury allegedly caused, in whole or in part, by relying on information contained in this book.

Questions regarding the content of this book should be addressed to
www.internationalmarine.com

Questions regarding the ordering of this book should be addressed to
The McGraw-Hill Companies
Customer Service Department
P.O. Box 547
Blacklick, OH 43004
Retail customers: 1-800-262-4729
Bookstores: 1-800-722-4726

To mariners everywhere who must balance the expectations of the shore against the demands of the sea

Contents

Prologue: The *Herald of Free Enterprise*   ix
Introduction: Why Bridge Resource Management?   1

Part I. Planning and Procedures   5

Chapter One. Standard Operating Procedures   7
Chapter Two. Passage Planning: Appraisal and Planning   17
Chapter Three. Implementing the Passage Plan: Execution, Conferring, and Monitoring   26
Chapter Four. Building a Passage Plan: Tactics and Tools   41

Part II. Situational Awareness and Human Factors   61

Chapter Five. Overreliance   63
Chapter Six. Distraction   72
Chapter Seven. Stress   81
Chapter Eight. Fatigue   88
Chapter Nine. Complacency   95
Chapter Ten. Transition   100

Part III. Human Interaction   109

Chapter Eleven. Communication   111
Chapter Twelve. Teams and Teamwork   123
Chapter Thirteen. Decision Making and Leadership   137
Chapter Fourteen. Human Error   145

Appendix: Sample Master-Pilot Information Exchange (MPX)   161
Sources   163
Acknowledgments   165
Index   167

Prologue

The *Herald of Free Enterprise*

On March 6, 1987, the modern cross-channel ferry *Herald of Free Enterprise* left its berth at Zeebrugge, Belgium, for the routine 70-mile run across the North Sea to Dover, England. There is, perhaps, no maritime endeavor more prosaic than that of a ferry. Four minutes after passing through the jetties, the *Herald* rolled over and sank with the loss of 188 lives.

The *Herald of Free Enterprise* sank because no one was aware its enormous bow doors had been left open, allowing the North Sea to wash in. Because of the tide that day, the bow had been ballasted down to facilitate loading, and the *Herald* remained trimmed by the head at departure. The assistant boatswain, whose job it was to close the bow doors, had fallen asleep in his bunk during the loading and did not awaken when called to get underway. His immediate supervisor, the boatswain, had been working near the bow doors prior to sailing and noticed that they were open and that nobody was standing by the controls, but he proceeded to his own station without giving it any more thought. He explained later, "It has never been part of my duties to close the doors or make sure anybody is there to close the doors."

The salvage operation in full swing after the *Herald of Free Enterprise* rolled on its side and sank. (Scaldis)

Written operational procedures stated that it was the duty of the loading officer to ensure the doors were secure when leaving port. This was not what usually took place, however. In practice, loading officers frequently left the area once they saw the assistant boatswain standing by the door controls. This enabled loading officers to get to their sailing stations so the ferry could sail on time. Sometimes the loading officer was the second officer (second mate); sometimes it was the first officer (chief or first mate). On this occasion it had been both. The first officer started the loading but the second officer relieved him early in the process. A little while later the first officer took over again without explanation. Subsequent interviews with both men suggested that on this particular day there was some tension between the two officers. When the first officer resumed the loading there was no clear transfer of responsibility; the two men did not meet face to face. The second officer heard the first officer giving loading orders over the radio and concluded that "the job had been taken away." When the loading was finished, the first officer headed for the bridge as required, but there was still no one at the door controls. The second officer, who by this time had left the scene, said later, "He took over as loading officer so I assumed he took the responsibilities that go with that job."

The first officer told investigators that he thought he had seen a man approaching the bow door controls but was then distracted. Whoever he thought he saw, it was not the assistant boatswain because the assistant boatswain was still asleep in his bunk, and there was no exchange with this other person related to getting the doors closed. It seems more likely that, in his haste, the first officer didn't notice, period.

Aggravating the situation was a recent memo from the operations manager of the ferry company that may have shaped the first officer's priorities: "Put pressure on the first officer [the chief mate] if you don't think he is moving fast enough.... Let's put the record straight. Sailing late out of Zeebrugge isn't on. It's 15 minutes early for us!"

Successive captains on many of the company's vessels failed to enforce the procedure that required the loading officer to ensure the bow doors were secure. This was poor leadership but it may be explained in part by the company's "Bridge and Navigation Procedures," which required the first officer to be on the bridge 15 minutes before sailing. Much of the time this was a physical impossibility when supervising the loading. It guaranteed that, wherever he was and whatever he was doing, the first officer would be rushing. The conflict had been pointed out to shoreside management in writing by at least one captain but the flawed procedure was allowed to stand, ensuring it would be violated on a regular basis. The problem with supervising the bow doors was a collective failure that contributed to a culture in which the violation of procedures became normal.

Why didn't the captain confirm that the doors were closed before sailing? A simple query to the first officer might have prevented all of this, but that sort of proactive communication was not standard. Furthermore, the company standing orders, as seen in the following excerpts, virtually guaranteed it never would be:

> Heads of Departments are to report to the Master . . . any deficiency which is likely to cause their department to be unready for sea [e.g., bow doors open] at the due sailing time.
> In the absence of any such report the Master will assume at the due sailing time that the vessel is ready for sea in all respects.

Assuming that all is well unless told otherwise is sometimes reasonable, but other times it is not. Situational awareness helps us discern the difference. On a vessel fitted with gaping doors, authorizing such an assumption was a written invitation to complacency, which the captain accepted. The captain was a company man. He operated within the system. Whatever imperfections he saw in it, he didn't see a need to do more. The system seemed to work.

Why didn't shoreside management employ more realistic procedures that would improve the operation rather than undermine it? One explanation is a lack of clear zones of responsibility at the management level. In trying to convey how responsibilities were designated at the management level one director provided an explanation that bordered on gibberish: "There were no written guidelines for any director. . . . It was more a question of duplication as a result of not knowing than missing gaps. We were a team who had grown together." Indeed.

Another explanation for shoreside management's failure to manage appears to have been a dysfunctional working relationship with shipboard leadership. For instance, repeated requests for door indicator lights went nowhere. Concerns were viewed as exaggerations. Six months before the incident one captain wrote, "Is the issue still being considered or has it been considered too difficult or expensive?" Apparently it was not too difficult or expensive, because within days of the accident, door indicator lights were installed throughout the fleet.

Logging the drafts of the *Herald of Free Enterprise* before sailing was legally required, but due to the docking configuration at Zeebrugge, it was generally not possible to read them. Trim is an important component of stability, and the need to alter the ferry's trim for loading meant that this information had special significance. More than one captain raised concerns about the inability to read draft marks but no effort was made to remedy the situation. The idea of installing high-capacity ballast pumps to enable a more rapid return to an even keel was rejected as too expensive. Instead, it became accepted practice to falsify the drafts so that they showed that the vessel was always on an even keel, an impossible feat in real life that no one cared to point out. As a result, the vessel routinely sailed with excessive trim, in violation of its certificate. This showed a lack of ethics both ashore and aboard, and further contributed to a culture of violation.

Over a period of five years before the accident no fewer than seven captains raised concerns about passenger overloading. They presented documentation showing that limits were routinely exceeded, sometimes by large amounts. The overloading was common knowledge aboard and also contributed to a culture of violation. Not until after the tragedy did management make a serious attempt to devise a system that would ensure its ferries did not carry more passengers than permitted. Once you stop playing by the rules, and get away with it, the tendency is to keep going.

High crew turnover and the shuffling of crew between vessels operating on different runs also meant that the personnel were in a constant state of transition. Concerns about this were raised by a senior captain, who pointed out that in the five-month period leading up to the accident the *Herald* had thirty-six different deck officers assigned to her. The captain wrote, "The result has been a serious loss in continuity. Shipboard maintenance, safety gear checks, crew training, and overall smooth running of the vessel have all suffered." Situational awareness can have only suffered too.

People say hindsight is 20/20, especially after neglecting to exercise foresight. This story has nothing to do with hindsight. Many people saw and articulated the problems. In particular, sailing with the bow doors open had been observed by many passengers over the years, with at least five documented instances of it prior to this tragedy.

Thankfully the *Herald of Free Enterprise* does not typify the maritime industry. It is, however, a stunning example of how human factors can leave an outwardly professional maritime operation hollow at the core. Though the *Herald* was not a watchkeeping failure per se, a great many of the breakdowns in this case fit squarely into the principles of bridge resource management (BRM): Standard operating procedures were absent, flawed, or disregarded. Situational awareness was undermined in every conceivable way. There was stress and distraction due to the pressures of the schedule. Fatigue was not a central issue, but the most important person, the assistant bosun, was asleep on the job, highlighting the difficulty of fitting adequate rest into the workday. Complacency was rampant: The accident investigation report described an organization infected by "a disease of sloppiness" and "staggering complacency." Transitions during the loading operation and due to high crew turnover impacted individual performance as well as teamwork, encouraging people to adopt minimalist attitudes toward their responsibilities. Human interactions were far from optimal. Communication breakdowns, both written and verbal, occurred at many points. Interactions between the two essential deck officers were strained, further undermining teamwork. The lack of leadership was widespread, contributing to a culture in which standards lost meaning. This multitude of human factors provided rich soil for errors to flourish and multiply, both detected and undetected. The tragedy of the *Herald of Free Enterprise* is an important case in the annals of BRM because it says so much about how technical prowess goes to pieces in the face of poor resource management.

The *Herald of Free Enterprise* may seem a peculiar case to open with. It was not a small ship, within the meaning of this book, and its sinking had virtually nothing to do with maintaining navigational control in the conventional watchkeeping sense. But wise mariners will take their lessons where they find them, and there are far too many here to look away. Also, the story of the *Herald* cannot be told without digging into the problems with shoreside management, a further remove from the bridge and watchkeeping. That this case holds lessons for shoreside management makes it all the better. We will revisit this story frequently for it is compelling proof of the need to apply the principles of BRM well beyond the wheelhouse door, an important concept to embrace from the outset.

Introduction

Why Bridge Resource Management?

As the year 2001 wound down and another Standards of Training, Certification, and Watchkeeping (STCW) compliance date drew near, mariners of all stripes packed into classrooms around the globe, needing bridge resource management (BRM, also known as bridge *team* management) certificates in order to keep their jobs. The mood was not pleasant. Many had decades of experience as captains and officers with immense responsibilities. They had seen things go wrong and experienced the consequences. They knew what they were doing. Now they were being ordered back to school to relearn what many felt either was obvious, couldn't be taught in a classroom, or wasn't worth knowing. Some veterans profited from the experience; others did not. As always, you get out what you put in. But BRM has never been about teaching old dogs new tricks. Like any deep change, BRM training is a long-term investment aimed at promoting a culture of safety across the maritime industry and across generations. You have to start somewhere, and you have to work at it for a long time.

Study after study confirms that between 75% and 90% of maritime accidents are caused in part or entirely by human factors. The underlying premise of BRM is that mariners are fallible in ways unrelated to technical proficiency. Of the catastrophes examined in this book, none were precipitated by mariners who didn't know how to navigate or drive a boat. Many of the main characters were very experienced. Though our technology has never been better, tragic accidents continue to occur, largely due to human factors. More and better equipment is always in the pipeline, but it is clear that technology is not the answer, or not the whole answer. Just as we cannot expect people to navigate who have never been shown how, we cannot expect people to be alert to the patterns of human error if the issues have never been raised.

If nobody minded the occasional oil spill or accidental death, BRM training might not be necessary. If ships were entirely automated, BRM training would certainly not be necessary. But that is not the world we live in, nor do we want to live in such a world. As professional mariners, we all make mistakes. When we suffer embarrassment as a result, that's a personal matter. But when avoidable accidents cause damage and harm others, it becomes a public matter, and that is what got us here. Mandatory BRM training maintains an industry-wide dialogue on the role of human factors in the safe operation of vessels. Of course, the best training in the world is worthless if it is not applied.

They say experience is the best teacher, and that is true. But the acquisition of experience can be harrowing. No one wants to serve as an example of what not to do, but it is one way that others learn. BRM training represents a commitment on a global scale to upgrade the human component of bridge watchkeeping in the interest of prevention rather than using ever more accidents as the training tool of choice.

BRM training presents special challenges compared to other new requirements. If you want a stronger ship, you simply increase scantlings. If you want to encourage the use of better navigational technology, you simply add the latest gadget. If you want to improve ocean survival rates, you place better survival equipment aboard. Regulatory agencies excel at this sort of upgrade because it is tangible, uniform, and easily verified. Modifying professional attitudes across vastly different maritime activities is a different kettle of fish.

Human Factors

BRM is about people, not boats. Although the discussion takes place in the context of vessel operations, in reality many of the issues are universal. BRM explores the benefits and the shortcomings of planning and standardized procedures. It examines well-known enemies of good performance such as fatigue, stress, distraction, complacency, the effects of transition, and overreliance on technology. Since mariners do not work in isolation, BRM looks at human interaction through communication, teamwork, and leadership. The ultimate goal is to prevent accidents; therefore we must also look at how seemingly insignificant human factors have the potential to compound and produce unthinkable yet avoidable accidents. BRM, like theater, is about human nature, with a vessel as the stage.

Many aspects of BRM are, in fact, traditional attributes of well-run vessels. All things being equal, navigating with a solid passage plan beats seat-of-the pants navigation. Proactive, unambiguous communication beats mind reading. Using position information from a variety of sources beats putting all your eggs in one basket. Standardized procedures for standard operations beat repeating painful learning curves. Standing watch with adequate rest beats standing watch exhausted. They say it's better to be lucky than good, and, believe me, there are days when that is true. But with so much on the line it is, in fact, better to be good than lucky. BRM is a way of packaging established truths so that all mariners can reap the benefits and apply that knowledge where it fits the situation.

In discussing the many elements of BRM it quickly becomes apparent that they are deeply entwined. We can't talk about planning and procedures without talking about human interaction because communication is integral to the success of plans and procedures. We can't talk about distraction without talking about fatigue or stress because each of these factors has the ability to aggravate the effects of the other two. We can't talk about teamwork without talking about leadership because they are hand in glove, and it's hard to talk about either of them without revisiting communication because that is how teams and leaders interact. Elements of BRM are not neatly divisible.

Small-Ship BRM

The mission of this book is to present BRM principles in the context of small ships, boats, or vessels. Technically speaking, this means vessels under 1,600 gross tons/3,000 international tons, sometimes referred to as *limited-tonnage* vessels. Regulatory cutoffs based on vessel size are notoriously awkward, making *small ships* an imperfect term. When we contemplate the spectrum of activities that takes place under that catchall term, it is evident that "small" is not so small, and the word "ship" is not always quite right either. Obviously, there is nothing diminutive about the responsibilities or the consequences associated with operating these vessels, and the sheer diversity of activities that falls into this category highlights the breadth of opportunity for the so-called limited license. Accuracy suffers from generalities, but if we are going to get anything else done, sometimes we have to tolerate a few.

Due to the range of small-ship operations, BRM principles cannot be applied the same way everywhere. On any given day, in any sphere of operations, some aspects of BRM will always be more applicable than others. Using the right tools for the situation is part of every job, and this is also true in applying BRM. Sometimes you must tailor BRM to suit the activity, but other times the entire point is to change the way the activity is conducted so that it is consistent with good BRM principles.

While common experiences bind all mariners at some level, differences do exist between the way BRM plays out in the world of limited-tonnage vessels as compared to deep-sea merchant ships or law enforcement or military vessels. This becomes evident as we explore the meanings of the bridge, the resources, and resource management.

The Bridge

BRM is traditionally focused on maintaining ironclad navigational control, with an emphasis on busy or confined waters. The basic idea is that a vessel should never be anywhere it isn't supposed to be, and mariners are responsible for seeing to that. The bridge is the physical location from which navigational control (the conn) is exerted. In the world of small ships this might be the wheelhouse, pilothouse, nav station, helm station or, yes, sometimes the bridge. On a sailing vessel it might be any of those, as well as the quarterdeck or the cockpit. The confines of a small wheelhouse may be such that a second person is almost a nuisance, even when trying to help. In such cases, the solo watchkeeper is the center of the wheelhouse universe, and all the tools should be arrayed within easy reach. But we also have megayachts, oceangoing tugs, research vessels, ferries, and offshore supply vessels that rival the largest vessels in sophistication and are equipped with ergonomically designed bridges to facilitate team operations.

But in reality, a lot of small-ship bridges are models of inconvenience: radar sets that require the operator to face aft, thus inviting disorientation; insufficient space to lay out an entire chart; laptop screens cluttered with multiple functions with an electronic chart running somewhere beneath; depth-sounders the operator cannot view from any natural position; noise, vibration, lousy visibility; do-it-yourself installations; and other limitations that just seem to come with the territory. On sailing vessels it is not uncommon to stand watch under the open sky, periodically ducking belowdecks to consult the navigational equipment. The "bridge" of a small vessel is anything but standard.

The Resources

Resources are anything that can help get the job done. They are commonly considered to include information, equipment, and people. Mariners use equipment and people to get information, but information also comes from countless conventional and unconventional sources: regulations, passage plans, stability letters, standardized procedures, free advice, the look of the sky, local knowledge, radio chatter, and the running lights of an approaching vessel. Mariners use this information to create and maintain *situational awareness*, an accurate interpretation of what is happening around them. Resources may be ship-based, like a spotlight used to pick out day marks, or they may be external, like a weather report or vessel traffic service (VTS).

Equipment on small vessels varies as much as the vessels themselves and the work that they do. A fast ferry may run twin gyrocompasses feeding an integrated bridge equipped with dual ARPAs (automatic radar plotting aids) and ECDIS (electronic chart display and information system) arrays, as well as night-vision apparatus for a bridge team of five. Another vessel may have a magnetic compass, an unstabilized radar, and a lead line. Equipment is not limited to bridge instruments: engines, anchors, and assist tugs are also essential parts of a watchkeeper's arsenal. It goes without saying that equipment is only as good as the knowledge possessed by those who use it.

People are the most versatile resource but also perhaps the most inconsistent. People have a unique capacity to gather information from any source, including personal experience, and use it to make fresh decisions. People, especially those on boats, are great problem solvers. But they possess different strengths and weaknesses, and in different measures. Even the same person isn't the same every day. People miscalculate, get tired, become distracted, or chafe at one another. They grow complacent, bored, and their wonderful minds wander. People are magnificent resources but they are far from perfect.

The most important people from a BRM standpoint are the crew. Other human resources include shoreside personnel, bridge tenders, lock tenders, pilots, VTS, and, of course, the person (or people) in the wheelhouse of the oncoming vessel. Sometimes the crew structure on small ships allows for a bona fide bridge team: a watch officer, a lookout, and perhaps even a third or fourth person contributing to the navigation effort as needed. Yet in much of the small-vessel world talk of bridge teams is met with incredulity: "What team?" Under the two-watch system a mate and captain operate almost as co-captains, each running the vessel for half the time. There may be a deckhand somewhere chipping paint, cleaning, or watching TV, but mostly it is a *team-of-one*. Frequently, a so-called bridge team isn't even on the so-called bridge, as when a deckhand is a thousand feet ahead, out on the front of a barge with a radio, relaying distances back to the mate as the tow gingerly enters a lock, or when an AB pilots a tow across a busy harbor because the view from the wheelhouse is obscured by a stack of containers.

One of the most useful resources is the past, if we can remember it. Mistakes can be instructive, and

The small figure on the bow of this barge is relaying information back to the obscured wheelhouse and is a critical "bridge" resource. Good captains and mates use all resources available and keep the lines of communication open.
(John Watson)

nothing concentrates the mind like a traumatic incident close to home, one you can picture happening to you. This book examines a number of incidents for their capacity to illuminate BRM principles and the consequences of seemingly minor human lapses. Every incident tells more than one story, though some illuminate certain themes particularly well. We look at these incidents, not to point fingers, but because it is not at all clear that we would do any better under the circumstances, and we want to do better. The human failings that lead to accidents are not exclusive to unworthy people, it's just that the people involved look bad in the aftermath. Bad things happen to good people, hardworking people, ordinary people.

Management

Resource management means resource *maximizing*—getting the best possible results from the available information, equipment, and people. Not all resources have the same value in a given situation, therefore resource management is often a prioritizing function: sometimes collision avoidance is more important than position fixing so we utilize our resources accordingly. Sometimes a radio conversation is helpful, but other times it is a distraction. Managing resources entails verifying information by cross-referencing: does a GPS position agree with ranges and bearings? Is that buoy on station? Though the bridge tender has hailed you through, do your eyes tell you that the bridge is fully raised? Resource management involves using the right information source for the job.

Managing resources includes managing people—teams. Teams offer the ability to distribute workload, which, in theory, means a team can do more than individuals functioning alone. But if a team is poorly managed—with poor communication, ill-defined responsibilities, and uncoordinated efforts—it can actually make matters worse. Completing almost any voyage is a team effort, no matter how rudimentary the operation. Resource management is how you determine that you need more resources. It is how you know when to call the captain, when to slow down, and when the schedule must bow to safety. Ultimately, only people can manage resources.

Casting Off

Occasionally we hear about how some aspect of BRM is at odds with how things are done in the "real world." This is surely true. But the real world changes under our feet. Things that were gospel at the outset of a career are history by the end, and not just the technology. Though we can learn from mistakes and the traditions of the past, BRM is not about the past, and sometimes not even the present. To study BRM is to look through a window at how things can be done better and, frankly, *will* be done in time. Just as the notion of what constitutes a "safe" operation evolves, so does our approach to resource management.

BRM has not brought an end to human error on the water. There will always be accidents, even catastrophes. Mariners deal with a tough environment, and there are always odds at work. Risk cannot be eliminated, but it can be managed with good training and good practices. Well-run vessels have accidents and close calls; poorly run vessels have more. Since human error plays a large role in this fact, it stands to reason that if you have a better understanding of normal human fallibility, you can improve your odds. BRM doesn't hold all the answers. In fact, it isn't hard to imagine a situation where an officer becomes so preoccupied with going by the BRM book that he or she fails to trust an instinct or some other source of wisdom at a critical moment. That would be bad. Nevertheless, BRM is a piece of a sensible strategy from which we all benefit.

Part 1

Planning and Procedures

THE VALUE OF GOOD PLANNING and procedures is self-evident. So why do so many accidents reveal a lack of good planning, procedures, or both?

One reason is that people often fail to envision the need for a plan or procedure until after an accident has occurred. This is the origin of many regulations and policies. To some extent failing to predict the future is forgivable, but that only gets you so far. No one could foresee that the assistant boatswain would fall asleep aboard the *Herald of Free Enterprise* and not turn up at his sailing station. But the possibility (and consequences) of the vessel sailing with bow doors open and trimmed by the head was eminently foreseeable, and better procedures could have prevented it. Though mariners cannot be expected to anticipate everything, they do have a professional obligation to at least *try*. A lack of professional imagination can be devastating.

Sometimes the best-laid plans and procedures do not produce the desired result because of miscommunication, misinterpretation, or reinterpretation. Somehow the intended meaning gets lost. This situation was evident aboard the *Herald of Free Enterprise* when, instead of standing by until the bow doors were secure, loading officers left the scene once they saw a person at the controls. This was accepted practice because it worked—usually. The best-laid plans are useless if they are not followed.

Another reason people fail to employ good planning and procedures is complacency. They have acquired a level of expertise and are aware of certain risks and hazards but they shield their eyes from them, telling themselves that it can't happen here, now, today, not to me.

The planning that we focus on in this section is passage planning. The procedures we are most interested in are those that enhance watchkeeping. What is good for watchkeeping is good for the ship.

A Shared Mental Model

Planning and procedures are fundamental instruments for creating a *mental model*, a predetermined idea of what should take place. You use the mental model to compare what *is* happening to what *should be* happening, and take appropriate action to keep your vessel and your world on track. What often separates a close call from a disaster is spotting a problem in time to do something about it. A mental model helps with this. A mental model that is overly rigid can be dangerous, but without mental preparation a complex task like running a vessel can only succeed through brilliant improvisation and luck. Both of these are good to have, but relying on them is thoroughly unprofessional. Therefore we create and follow plans and procedures to help us, and those around us, uphold our responsibilities to operate as safely as we know how.

Since most mariners do not run their boats alone, it becomes necessary to *share* the mental model so that everyone is pulling together and in the same direction. A shared mental model helps the greenhorn deckhand and the most exalted captain act in tandem. When several people are working as a navigational team, a shared mental model makes that team stronger by providing a basis for prioritizing effort and information. Nobody sees everything, but everybody sees something. That deckhand with a radio at the head of a 1,000-foot string of barges can give better information to the watch officer back in the wheelhouse if they share the same mental model. And even the solitary watchkeeper must go off watch at some point. A shared mental model based on a detailed passage plan and well-designed operating procedures can help the oncoming mate carry the voyage forward using agreed parameters.

The goal of a shared mental model is not to get the team to agree for the sake of agreement. The benefits of a shared mental model must be balanced against the possibility of a team succumbing to *groupthink*, that is, the pursuit of conformity and cohesion at the expense of independent thinking. In a team setting, the shared mental model exists to help the left hand and the right hand work together. A shared mental model is *not* navigation by committee, but it is collaboration.

Single-Point Errors

Single-point errors occur when the success of an operation hinges on a single resource—equipment, information, or a person—and that resource fails. Redundancy is a hallmark of all safe systems, and it helps prevent single-point errors. Sometimes it is just a matter of having enough hands. But without a shared mental model, an army of people can do little more than trample down grass. When watchkeeping is a solo performance, as it often is aboard smaller vessels, the situation is particularly vulnerable to single-point errors of judgment and calculation. In fact, such errors are only a matter of time. To compensate, solo watchkeepers, even experienced ones, need every available tool to steer the odds in their favor. Good procedures and a detailed passage plan are just such tools.

Chapter One

Standard Operating Procedures

If there is a right way to do something, sure as the sun rises in the east someone will find a wrong way to do it. For that reason even the simplest, most informally run vessels have routines. You check the oil before starting the engine. You look over the stern to make sure nothing will foul the prop. You test the gear ahead and astern. You disconnect shore utilities before casting off. You make regular rounds checking bilges, machinery spaces, and other compartments for signs of fire, flooding, or things coming adrift. These practices are rooted in experience, most of it bad: seized engines, fouled props, torn shore connections, and other fiascos.

Unlike an informal routine that relies on personal familiarity, word of mouth, and accurate recall, *standard operating procedures* (SOPs), or just *standardized procedures*, define the way things should be done in a professional environment. In the marine industry, standardization begins with terminology. Calling shipboard objects by their proper names enables crew to get their jobs done with less confusion than if they referred to everything as "that thing over there." No one disputes the value of standard terminology, although regional or sector-specific variations do exist.

SOPs are a mechanism not only for achieving routine efficiency but also for preventing accidents, small and large. All industries use SOPs, because once a dependable way of doing something is understood, people can get consistent, repeatable results by doing it the same way every time. Standardized procedures bring predictability to workers' actions while allowing them to anticipate the actions of others. That doesn't mean things will never go wrong, but good standardized procedures are a way of preventing avoidable mistakes and are part of every successful organization.

Like terminology, SOPs are a form of communication. An SOP spells out the way to go about doing something. It should not change arbitrarily, or with the comings and goings of the crew or even the people in charge. Yes, there is more than one way to skin a cat, and adaptability is the mark of a good shipmate, but when the same thing is done more than one way, then it is not standardized.

Though SOPs bring uniformity to work, they are not static; SOPs evolve. That is why handbooks and manuals must be updated from time to time, and crew must be retrained. There are occasions when SOPs must be overridden or modified temporarily. When this happens, you must proceed with caution, recognizing that the operation is not going as usually performed—thus it is a departure from procedure. When operations depart from the prescribed procedure on a regular basis, the entire enterprise is exposed. We must either change the procedure to fit the practice, or change the practice to fit the procedure. Living with a discrepancy between practice and written procedure is a clear signal that an accident is in the making, though years may pass before it happens, as with the *Herald of Free Enterprise*.

For SOPs to be effective they must meet three criteria. First, they must be professionally sound from both a regulatory standpoint and in terms of best practices. If procedures don't make good professional sense they won't prevent accidents and they won't be followed. Professionally sound procedures are realistic, achievable, and supported by adequate resources. As we saw with the *Herald of Free Enterprise*, the procedures and directives that required an officer to personally confirm that the loading doors were secure while also getting to the bridge in time to assist with getting underway were professionally unsound. Procedures that called for minimizing trim before sailing when the ballast system lacked the capacity to accomplish that were professionally unsound. Procedures that instructed the captain to assume the bow doors were closed unless told otherwise were particularly unsound for a vessel with such perilous vulnerabilities. In each case, the resources—people, equipment, and information—were either inadequate or poorly managed, making the procedures unrealistic and unsound. Better procedures would have erected stronger barriers to the vessel getting underway before it was ready for sea.

Second, a standardized procedure must mean the same thing to everyone. This understanding usually starts by putting it in writing. Procedures transmitted by word of mouth are inevitably miscommunicated and misunderstood over time. The language must be clear and unambiguous, not because mariners are averse to shades of gray, but because they are as susceptible as any other group of people to arriving at the most convenient interpretation possible. Clarity is not always achieved on the first attempt, so developing good procedures requires follow-up and revision. One test of clarity is if a procedure is readily understood by a new crew member without verbal qualifiers, modifications, or amendments. How many times have you been shown procedures only to have someone dismiss some portion, saying, "Don't worry about that part. We never do that"? Even clearly written procedures may be reinterpreted over time. Aboard the *Herald*, ensuring that the bow doors were secure came to mean ensuring someone was at the controls. Therefore, every organization must commit to making sure that crew understand the meaning and the intent of its procedures. This is known as training, and it costs money. Among other things, SOPs are a means for establishing accountability, but to hold people accountable, the procedures must be accessible in an unchanging form to those who are expected to follow them. For this reason many SOPs are posted, or placed in binders at the nav station, or appear in crew handbooks and vessel familiarization materials.

Third, the procedures must be followed. A procedure that meets the first two criteria is useless if people don't adhere to it. If the first two criteria are met, however, people are far more likely to follow the procedures that result. Unfortunately, there is always someone who doesn't care, who is complacent, or who by nature is a risk taker. Because of this element, and the paramount obligation to keep the ship and the people in it safe, it is sometimes necessary to enforce procedures. The only way so many critical procedures were violated on *Herald* on such a regular basis was through a lack of enforcement.

The *Andrew J. Barberi*, 2003

The *Andrew J. Barberi* was a 310-foot passenger ferry built to make the 22-minute run between Manhattan and Staten Island, New York. Capable of carrying 6,000 passengers, it was part of a fleet run by the New York City Department of Transportation. The ferry system in New York City is substantial, and in 2003 it had a century of operating experience. It had 450 employees, at least five vessels operating at any time, an annual operating budget of nearly half a billion dollars, and transported some 19 million passengers a year. An organization of this complexity and magnitude has particular need for standardized procedures.

The *Barberi* was configured with a bridge at each end so it could be operated in either direction, but only one bridge was operational at any given time. Navigation was handled by a captain and an assistant captain supported by an engine room staff, seven deckhands, and two mates who oversaw loading and unloading. The captain had been with the ferry service for thirteen years, and the assistant captain for eighteen years. They had both received high marks on performance evaluations.

For standard operating procedures, the *Andrew J. Barberi* had a seven-and-a-half-page document describing the duties of the captains, assistant captains, mates, seven different deckhand assignments, and the bathroom attendants during loading, unloading, and underway operations for three different classes of vessel. A paragraph each was devoted to the captain and assistant captain duties. Here are some key provisions:

Captain

Upon docking the Captain will be in the inshore pilothouse insuring the aprons . . . are in the correct position to receive the boat and the slip is in otherwise safe condition to dock.
Upon undocking the Captain will be in the offshore pilothouse to receive the signal from the Assistant Captain that the boat is let go.
The Captain (taking full responsibility) may allow the Assistant Captain to make landings at Whitehall [the Manhattan end], maneuver in and out of tie up and fueling slips or get the boat underway.

Assistant Captain

Assists the Captain in the operation and navigation of the ferry.
After the boat is secured . . . observes the off loading and loading of the vessel in order to react to any problems that may arise during this time.
As soon as the boat is loaded . . . and safe to depart, the Assistant Captain will signal the Captain [in the offshore pilothouse] and transfer the controls.
After the boat clears the slip, proceeds to the operating Pilothouse. Makes log entries.

Though it is not plainly stated, the implication is that the captain and assistant captain would both be on duty in the operational pilothouse for the transit as soon as possible after getting underway. The docking duties could be interpreted a number of different ways, none of which is easily squared with the actual long-standing practice, which was that the captain docked at the Manhattan end (Whitehall) and the assistant captain docked at the Staten Island end, once the captain deemed the assistant captain skilled enough.

In addition to these procedures, a *bridge deckhand* was assigned to report to the operational pilothouse shortly before each arrival to make an arrival announcement over the public address system. Though not specified in the standard operating procedures, it was customary to have a deckhand on the bridge for each transit acting as lookout until relieved by the bridge deckhand.

On a normal run to Staten Island, a number of interrelated events would occur around the time the ferry passed the KV buoy. First, the vessel would begin to slow in preparation for docking. For both crew and passengers, the slowing of the engine was an audible cue that they were approaching their destination. Thus alerted, the designated bridge deckhand would go to the operational pilothouse to make the arrival announcement. When the bridge deckhand arrived in the pilothouse, the lookout was free to go to his or her docking station. As long as this sequence played out, the assistant captain was never alone on the bridge even if the captain was elsewhere. But this sequence did not always play out. If the vessel was crowded, the rush of passengers could prevent a lookout from reaching the docking station in time. If the lookout had to beat the crowd to get where he or she was supposed to be, that meant leaving early. The need to be in two places at once was unrealistic and is uncannily reminiscent of what happened aboard the *Herald of Free Enterprise*. Sometimes the bridge deckhand found it easier to make the arrival announcement from the nonoperational bridge, and did so. Sometimes both captains

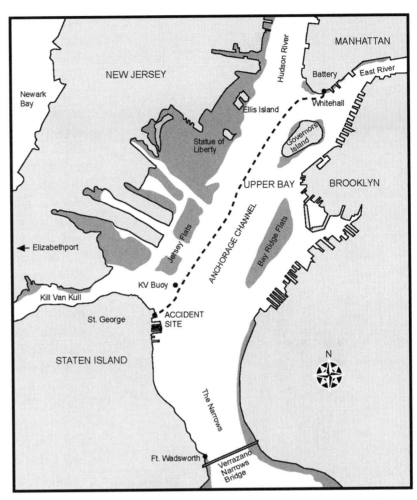

The *Andrew J. Barberi*'s regular route between the Battery (Whitehall) on Manhattan Island and Staten Island. (National Transportation Safety Board)

were not on the bridge for the transit. There were elements of a routine, but it was certainly not standard.

On the afternoon of October 15, 2003, the assistant captain had the conn as the ferry approached the Staten Island dock. A deckhand was seated beside him keeping lookout. One of the mates was reading a newspaper in the back of the pilothouse, but he had no navigational duties and was not part of the bridge team. From his seated position he did not have a view out the window. The captain was at the far end of the vessel in the other pilothouse doing paperwork.

On this particular day, there was a door with a faulty latch on the passenger deck that had been lashed shut to prevent it from slamming. The lookout left the pilothouse before the bridge deckhand arrived to unlash the door so passengers could disembark. Thus what was set up to be a three-person bridge team was now a one-person bridge team. That one person, the assistant captain, apparently blacked

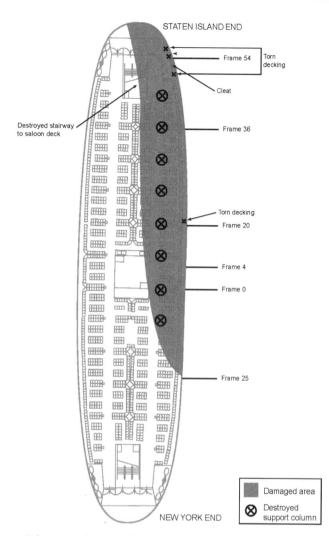

Schematic showing the damage to the *Andrew J. Barberi*. (National Transportation Safety Board)

out while standing at the controls. He did not collapse nor did he ease off on the throttles at the KV buoy. Without the change of engine speed the bridge deckhand didn't know it was time to go to the bridge. The arrival announcement was never made so the passengers had no warning that the ferry was approaching the terminal. The mate, a few feet away, was absorbed in his newspaper and had no view of the approach. Of a deck crew of eleven, only one deckhand was aware that the vessel was approaching the dock. Everyone else was taken completely by surprise when the ferry slammed into the dock at 15 knots, ripping a 210-foot gash down the side of the vessel.

Ten people were killed outright, and one died later. Seventy people were injured, including some who were severely maimed. Damage to the vessel and the dock totaled $8.4 million, and the city paid over $67 million to settle lawsuits. As it turned out, the assistant captain, with the complicity of a doctor, had concealed a medical condition from the Coast Guard, who issued his license, because he was concerned he might lose his job. He was taking a number of medications for a variety of ailments, and the loss of cognitive ability was a known potential side effect. After an attempted suicide later on the day of the accident, he served fifteen months in prison. The captain was fired. The director of ferry operations was sent to prison for not enforcing the rule requiring two pilots on the bridge. One port captain received two years' probation for making false statements to investigators. Many lives were shattered.

The *Andrew J. Barberi*'s high-speed collision with the pier at Staten Island killed ten people instantly and ripped a 210-foot gash in the side of the ferry. (Associated Press)

While the precise nature and timing of the assistant captain's incapacity was unpredictable, whoever had the idea of requiring two licensed officers and a deckhand on the bridge was clearly onto something. It wasn't necessary to foresee how this particular tragedy would unfold in order to understand what was at stake and how to prevent a single-point error in the pilothouse. The procedure that could have best avoided this incident was the one requiring the captain and assistant captain to stand watch together. As it turned out, this was routinely disregarded. Many crew were unaware of any such requirement, and the captain's absence from the pilothouse prior to the accident was considered unremarkable. When investigators asked the two port captains about procedures governing the bridge team, they gave two different answers. One port captain said that the two officers were required to be on the bridge together once they were away from the dock, but the other said it was up to the captain: "If he doesn't want to be in the pilothouse, if he has to go do something, he can do it, as long as he knows a qualified person is in the pilothouse." According to the director of operations, to whom the two port captains reported, there was no single, explicit directive for how the captain and assistant captain were to interact. He said, "You have to kind of put [the procedures] together.... You see, they are to be together. In weather, absolutely, positively together all the time." These procedures meant different things to different people, and some people didn't know of their existence. When the meaning of a procedure is *between* the lines rather than *in* the lines, it opens the door to interpretation, and that was part of the problem aboard the *Barberi*.

Not only did the written SOPs contain ambiguities, it became apparent in the aftermath that they seldom saw the light of day. They were undated, making it impossible to know when they had been written or if they were current. They had not been disseminated systematically, which defeated the opportunities for standardization. Training and orientation for new crew were handled verbally as things came up. Management directives also were communicated verbally, and there was no system to document who had been told what. This lack of uniformity became both a symptom and a cause of a lax operating environment. As standards diverged, enforcement quite naturally fell by the wayside. Small wonder that the original concept of having a bridge team dissipated over time.

Many people contributed to the creation of a professional environment in which it became acceptable to not follow procedures. Following clear, professionally sound, and realistic procedures was the ounce of prevention that was lacking here. Unfortunately, there is no pound of cure for what happened.

Watchkeeping Procedures

Standardized procedures come in many forms and carry different amounts of weight. Some reflect the idiosyncrasies of the vessel and operation, such as the station bill, but others are standard throughout the maritime world, such as logging in and checking that the running lights are on when taking the watch. SOPs may be established through regulation, company policy, vessel policy, or best practices. The most important SOPs are reserved for high-stakes operations that are prone to human error. The ones we are most interested in are those which intersect with watchkeeping duties.

Regulations

Regulations make things regular. They standardize procedures in a way that has legal significance. The captain of the *Herald of Free Enterprise* was required by law to log his drafts marks before sailing. Since trim can have serious ramifications for stability, requiring him to monitor draft was a good way to keep the vessel seaworthy. But there was no convenient way to get this information, so the logbook routinely was falsified and the vessel sailed trimmed by the head with bow doors open, but on schedule.

Many of the most important SOPs originate in regulations established at the local, national, or international level. Pilotage requirements, for instance, are often established by the port, but they have important ramifications for the composition of the bridge team and how the ship is navigated. Prearrival and predeparture equipment checks are commonsense precautions, and regulations frequently codify these checks in great detail. Another example of regulatory SOPs are the COLREGS (International Regulations for Preventing Collisions at Sea), also known as the Navigation Rules or the Rules of the Road. These rules standardize conduct between vessels in order to prevent collisions, thereby bringing a modicum of predictability to the actions of vessels all over the world.

Just about every collision involves failure to follow these procedures.

The Standards of Training, Certification, and Watchkeeping (STCW) Convention is an international agreement that imposes standards for seagoing personnel. Among these is BRM training because so many navigational accidents are traceable to a departure from good BRM practices. New regulatory SOPs are in the works all the time, usually in response to an accident.

Since a license is portable, a mariner may work on a wide variety of vessels in the course of a career, and come under a variety of regulatory SOPs. For example, a Certificate of Inspection (COI) spells out SOPs for manning, safety equipment, and other operational requirements for a vessel specific. Another example is a stability letter. While some of its stipulations are standard, such as keeping tank cross-connections closed while underway, others are vessel specific. Taken together, the COI, the stability letter, and other official instructions are the source of many of the most fundamental shipboard SOPs. Regardless of whether an officer reads, understands, agrees, or complies with these procedures, he or she is still responsible for following them. And if the captain and officers do not follow the procedures, who will?

One of the most far-reaching regulatory sources of standardized procedures is the International Management Code for the Safe Operation of Ships and for Pollution Prevention, commonly known as the International Safety Management (ISM) Code. This initiative was spurred in part by the *Herald of Free Enterprise* disaster, which revealed numerous organizational and procedural flaws. The ISM Code went into force in 1998 and has since been extended to virtually all commercial vessels on international routes. The purpose of the ISM Code is to compel vessel operators to develop a safety management system (SMS) for their vessels. An SMS is a comprehensive system for minimizing risk and fostering a culture of safety through standardized procedures, checklists, and safety audits. Every vessel coming under the ISM Code must have its own safety management system. Following is an outline of the goals of an SMS:

- Provide for safe practices in vessel operation and a safe working environment on board the type of vessel the system is developed for.
- Establish and implement safeguards against all identified risks.
- Establish and implement actions to continuously improve the safety management skills of personnel ashore and aboard vessels, including preparation for emergencies related to both safety and environmental protection.
- Ensure compliance with mandatory rules and regulations.

The ISM Code did not apply to the *Andrew J. Barberi* because it operated exclusively on domestic waters. However, many organizations falling outside the ISM Code have voluntarily adopted the same standards or developed parallel programs. This includes commercial operators on domestic waters as well as large private yachts operating internationally. While higher insurance premiums and the specter of more regulation often motivate voluntary compliance at the corporate level, at the operational level it is important for individuals to embrace the notion that just because a particular procedure is not required does not mean it isn't a good idea.

Internal SOPs

Standardized procedures from the regulatory world shape the way crew go about business aboard, including standing watch. But many critical operational details are left to the crew to figure out. Some procedures, such as standing orders, are commonplace, yet differ from ship to ship and captain to captain. Some organizations provide SOPs for the crew to read and sign. These are generally based on long operating experience but may contain ideas imported from elsewhere. Other SOPs are generated by people aboard who have recognized a need. Anyone can come up with a better procedure *after* an accident. A greater challenge is to create procedures that anticipate which things are likeliest to go wrong and how they will go wrong. Without realizing it, many people owe their careers to homemade, proactive procedures that quietly prevent accidents year after year by molding behavior and reinforcing good habits. Internal procedures vary in content and in scope but a few deserve special attention, partly because they are so widespread.

Standing Orders

Standing orders stand, come what may. They put key operational parameters in writing for watchkeepers. Standing orders establish a two-way street of obligations between watch officers and the captain. The mate is often required to sign the standing orders, like a contract. Among other things, they stipulate thresholds for calling the captain, including the ubiquitous catchall, "When in doubt." Standing orders put a mate on notice as to the expectations for standing watch, and they make clear that, under certain circumstances, calling the captain is an order, not a suggestion.

Calling the captain must be done in time for the captain's presence to make a difference. As one captain said, "If there's going to be a collision, I want to be a participant, not just a spectator." Most captains sleep better if they know they will be called in time. In return, the captain has an obligation to provide appropriate support when called upon. On vessels where the captain is one of only two watchkeepers, time off watch is sacrosanct. But if you can't call the captain when in doubt, who can you call? Those situations call for highly experienced mates who seldom use the off-watch captain as a resource.

Because standing orders are always in effect, they have to accommodate a wide variety of situations, yet they may also contain very specific parameters, such as minimum CPA (closest point of approach) with traffic and hazards. Most standing orders lay out expectations for monitoring traffic, communication equipment, and the ship's position. Some common circumstances requiring notification of the captain/master are the following:

- The vessel is in danger, either underway, at anchor, or alongside.
- There is an equipment failure or malfunction.
- There is an injury or illness.
- Visibility falls below XX miles.
- You are shorthanded or require assistance.
- Anticipated navigational benchmarks (sighting land, navigational aids, obtaining soundings) are not met.
- Unexpected sightings of land, navigational marks, or changes in soundings are encountered.
- Progress is impeded, the vessel is not handling as it should, or you are unable to maintain course or speed for any reason.
- When in doubt.

These instructions can be expanded, reconfigured, or refined to suit many situations, yet it is remarkable how many circumstances a few good directives can cover. When standing orders are generalized this way, it is incumbent upon the watch officer to mentally sketch out the potential situations to which the standing orders apply. That's one way to pass the time on watch.

Night Orders

Night orders are not standardized procedures in the usual sense, as they lack the permanence of standing orders. But they are an important mechanism for guiding a watchkeeper's decisions and must be routinely coordinated with standardized procedures. Night orders temporarily modify or augment the standing orders with additional details appropriate to the circumstances. For instance, with poor visibility night orders may stipulate larger CPAs than usual. Or, due to the activities of recreational vessels, they may permit a smaller CPA than usual with some types of traffic. Or they may specify a certain speed to maintain so as to make a critical waypoint by a certain time. Whatever the night orders stipulate, if they conflict with other SOPs, then the watchkeeper must be clear as to which takes precedence.

Night orders are customarily signed to affirm that they have been read and understood. Both standing orders and night orders carry as much authority as a verbal order given face to face. In the event of an accident, standing orders and night orders, like other SOPs, can be used to exonerate someone or establish fault.

Watch Conditions

In the normal course of a voyage, there is an ebb and flow to the intensity of the navigational demands. Sometimes things run smoothly to the point of boredom; other times you know you are earning your pay. When the demands of a situation ramp up, the standard watchkeeping arrangement, whatever it may be, may no longer be adequate. If situational awareness is unable to keep pace with events, there is a resource imbalance and elevated risk. One way to reestablish equilibrium between resources and demands is to slow down. Another way is to bring more resources to bear. Watch conditions expand upon standing orders and night orders to define the composition of the bridge team for certain operational conditions. Watch conditions are essentially a resource plan for a range of navigational situations.

The circumstances that trigger a change in watch conditions vary, but you can imagine many of them.

WATCH CONDITION	SHIP'S EXTERNAL ENVIRONMENT		
	VISIBILITY	WATERWAY	TRAFFIC
1	Unrestricted	Offshore Waters	Light
2	Restricted	Restricted	Moderate
3	Restricted	Restricted	Heavy
4	Restricted	Pilotage Waters	Heavy

A variety of factors influence the way watch conditions are structured. Watch conditions are often set by company policy and augmented by the ship's master. (Adapted from the Maritime Institute of Technology and Graduate Studies and Pacific Maritime Institute's Bridge Resource Management Course, 2002)

Some watch conditions are driven by external factors (fog, darkness, traffic), but others may be driven by the nature of the vessel or operation. A sailing vessel entering port may require one officer to direct sail handling while the captain takes the conn and yet another officer continues to navigate. Some vessels require additional resources for docking or managing a barge, but others do not. A failure of critical navigational equipment could trigger an elevated watch condition if it requires reverting to more time-consuming methods of navigation. Watch conditions mainly pertain to the bridge team, but they can involve other crew members. For instance, it is not uncommon to designate that the engineer be standing by in the engine room for critical maneuvers. Here are some examples of circumstances that might trigger a higher watch condition:

- Landfall
- Entering confined waters
- Maneuvering for docks, locks, or anchorages
- Reduced visibility
- Heavy traffic
- Heavy weather
- Loss of critical equipment

Watch conditions can include specific bridge assignments, such as collision avoidance, position fixing, communications, logkeeping, and lookout. When a bridge team is functioning well, there is some natural overlap in awareness, but allocating duties helps minimize blind spots and avoid wasteful duplication of effort.

One benefit of watch conditions is that they are proactive. By anticipating areas of high and low watchkeeping intensity and assigning resources accordingly, watch conditions take some of the guesswork out of managing bridge resources and help reduce the chance that a watchkeeper will seek backup that is too little, too late. Imagine coming into an unfamiliar port at night with multiple targets, isolated hazards, complex traffic patterns, and current. You're holding it together pretty well but constant radio chatter on multiple channels is making it hard to understand some of what is happening around you. You could tough it out and hope for the best, but a better choice would be to get some support. The establishment of watch conditions can help prevent situations in which watchkeepers become so preoccupied with what they are doing that they do not realize they are overtasked.

BRIDGE TEAM DUTIES BY WATCH CONDITION							
WATCH CONDITION	BRIDGE TEAM DUTY						
	CONN	COLLISION AVOIDANCE	RADIO COMMUNICATIONS	NAVIGATION	OTHER DUTIES	HELM	LOOKOUT
1	Watch Officer					AB	
2	Master	Watch Officer				AB	AB/OS
3	Master	Watch Officer(s)				AB	AB/OS
4	Pilot	Master		Watch Officer		AB	AB/OS

An example of one shipping company's classification of bridge team duties based on various watch conditions. (Adapted from the Maritime Institute of Technology and Graduate Studies and Pacific Maritime Institute's Bridge Resource Management Course, 2002)

Also, sometimes an officer is reluctant to seek assistance because it might reflect poorly on his or her abilities; it might be interpreted as not being up to the job, not filling the shoes, or not having the right stuff. This is an unfortunate and dangerous mental exercise that has played out in the minds of many officers at one time or another. Watch conditions can help relieve some of the decision-making burden around this.

Watch conditions are not applicable to every operation. Sometimes the nature of the work doesn't require them or the resources are simply not available. Overload must be prevented in other ways. However, watch conditions are an established bridge team management tool. Even if formalized watch conditions are not implemented, the concept of matching resources to changing situations is fundamental to bridge resource management. Because crew do not always recognize when situational awareness is slipping away, any mechanism that helps them stay on top of the situation has value. Watch conditions help provide an organized response to what your gut may already be telling you.

Checklists

Experience is a great teacher, and the best cure for inattentiveness is the experience of neglecting a step in a procedure that permits no omissions. An example of this might be neglecting to switch over to manual steering upon entering a busy stretch of water and then losing precious seconds fumbling with a recently installed autopilot as a target makes an unanticipated course change in your direction. Or failing to confirm your position at the change of the watch and steering for what you think is the next buoy but which in fact is *two* buoys away, and you have just left the channel on your way to it. Provided you and your career survive such moments, you will likely never make the exact same mistake again. A procedure requiring manual steering at certain junctures or confirming your position by more than one source cannot eliminate negligence, but it can create a clearer pathway to a safe operation.

The more complex the operation and the greater the demands it places on individuals, the more likely it is that something will be overlooked. Setting up certain SOPs in the form of a checklist is a established remedy for this. Here are some examples of common standardized shipboard procedures that lend themselves to checklist format:

- Pre-underway equipment tests
- Prearrival equipment tests
- Noontime equipment tests
- Fire and security rounds
- Transferring the watch
- Fuel transfers
- Confined space entry
- Regular inspections of critical equipment

A checklist serves as an unchanging memory prompt when detail and sequence are critical. No matter how many times you have performed a procedure, there will always be days when distraction, stress, and fatigue can affect performance. Because checklists standardize our actions, they also help keep us on top of things on days when complacency is running high and some part of the brain is contemplating cutting a corner. This temptation is especially likely in work that involves a high degree of repetition. And note that the second page of every checklist should start with a reminder to check that the actions listed on the first page have been completed!

Checklists do have their limitations, though. They can be a source of complacency themselves if people run through them without really thinking about the significance of what they are checking. What if something isn't on the checklist? Crew still need to use their *professional peripheral vision* to notice things that aren't right. Checklists are a proven mechanism for doing things the same, right way, but they are no substitute for independent thinking.

The pervasiveness of checklists in the workplace also can be a drawback if people come to perceive them as an exercise in paperwork, rather than a means of getting things right. The point of checklists is to do things properly, not merely to generate pieces of paper. However, don't overlook the fact that checklists can serve as a paper trail to show whether procedures were followed.

Certain professional habits don't lend themselves to checklists, and that's just as well. The ability to notice a wind shift or a change in the vessel's motion, perceive the potential ramifications of those changes, and make intelligent decisions as a result doesn't come from a checklist. That process comes from the little masterpiece called "experience" that every good mariner is painting in his or her mind all the time. But if a checklist helps prompt good watchkeeping practices or reveals a critical shipboard deficiency in a timely fashion, some accidents won't happen, and that is good.

*****

Standardized procedures have played a part in the long-standing trend toward more paperwork, and nobody is drawn to the water for the love of paperwork. When a procedure appears to have been drafted by legal counsel or by a regulator in an office far from navigable waters it can be frustrating for those who must live with it. Procedures that emerge from the aftermath of an accident often have a knee-jerk quality that can make competent mariners feel hamstrung.

Mariners are often quick to rail against procedures they find objectionable or burdensome, but the fact remains that a world without sound SOPs is more hazardous than a world with them. The survivors of the *Andrew J. Barberi*, the *Herald of Free Enterprise*, and many other incidents can attest to this. Most SOPs, well designed or not, have a rational underpinning either in common sense or, unfortunately, in an accident. The best procedures anticipate human failings such as memory lapses, haste, and complacency. They balance hands-on attention to detail with a big-picture view of the overall goals. Standardized procedures won't prevent all accidents, but they are an important line of defense.

Chapter Two

Passage Planning: Appraisal and Planning

Plan your work and work your plan. This is one of the more basic pearls of wisdom out there. Passage planning is the process of marshaling, sifting, and organizing information (charts, publications, protocols, regulations, local knowledge, etc.) to produce a documented plan and a carefully considered route for a given transit. This information is used and consulted throughout the voyage to keep the vessel in safe water and out of trouble. The main focus of passage planning is on maintaining navigational control, but there are other elements. Together with sound standardized procedures, passage planning can do much to preserve a seasoned mariner's good reputation while giving a young mariner the chance to earn one.

Many aspects of passage planning fall within the realm of common sense, which is why all mariners, even recreational boaters, do it reflexively. We find ourselves facing the same basic questions before getting underway: Where are we going? How far is it? How long will it take? Is the water deep enough? Is the vessel up to it? What are the hazards? How will we overcome them? We pose such questions without prompting because we want to arrive safely, and we know there are no guarantees. At one level, passage planning is a form of risk assessment and risk management. In a professional setting, managing risk must be more than the chance linking up of good ideas, sticky-note reminders, and passing conversations. Passage planning at the professional level is a structured approach to thinking ahead. It should be deliberate, systematic, and transparent.

Like all planning, building a passage plan dials you into the future. It strengthens your powers of foresight and minimizes the unexpected by creating an opportunity to consider potential problems in advance. A passage plan enhances situational awareness by giving you a solid basis for comparing what *is* happening to what *should be* happening, a critical real-time function of watchkeeping. This is especially helpful when the demands of the watch are high or fatigue and distraction are working against you. By addressing much of the navigational detail beforehand, the watch officer is free to focus on the truly unforeseeable. It is infinitely easier to cope with traffic if you are not also attempting to lay track lines, read the *Coast Pilot*, and correct charts between glances at the radar.

A passage plan is a communication device that facilitates the development of the shared mental model. Too often capable crew have been rendered ineffective because there was no mechanism for systematically sharing information. A documented plan and a detailed route make it possible for crew to be literally on the same page. The transparency of a passage plan clarifies goals, tactics, and tolerances while providing a tangible point of reference for discussion, review, or modification. By detailing expectations, the plan provides more comprehensive guidance than standing orders or night orders, and this benefits the captain when he or she is below.

With a detailed plan in hand, a watchkeeper is less likely to eyeball the situation and hope for the best. This helps manage complacency. A passage plan, like standardized procedures, provides a pathway for good decision making. This may be especially valuable for those with less experience.

Unlike the electronic components in which we place so much faith, passage planning is essentially low-tech. Its effectiveness, however, is directly related to the way people approach it. If the plan becomes a perfunctory exercise to humor the home office, or if it encourages navigators to view themselves as a train on a track, there are likely to be problems. For this reason a passage plan should not be viewed as a straitjacket. There may be good reasons for departing from it, which is why *contingency planning*, the process of expecting the unexpected, is part of it.

Berth-to-berth passage planning is an expectation of our times, and a major cornerstone of BRM. Both the International Convention for the Safety of Life at Sea (SOLAS) and the STCW mandate it. Passage planning is not something to do if you get around to it; it is part of what a deck officer is paid to do and the workday must leave time for it.

The IMO (International Maritime Organization) has established a conceptual framework for passage planning that comprises four stages: appraisal, planning, execution, and monitoring. We will look at the first two stages in this chapter and the second two in Chapter 3, along with a fifth element, conferring, which is often overlooked. Through the prism of case studies it soon becomes apparent how a more methodical approach to passage planning could have helped seasoned and well-regarded professionals avoid serious accidents. In Chapter 4 we learn tactics for safe route planning and look more closely at the components of a passage plan.

Appraisal

The appraisal stage of passage planning is essentially a feasibility study and a risk assessment rolled into one. The planner often starts by scrutinizing the proposed route and the destination to determine whether it is even realistic. He or she uses the physical dimensions of the vessel to ascertain the smallest under-keel and overhead clearances that could possibly be encountered, including the effects of tide, lunar phase, squat, water density, and other potential changes to draft. The planner must also consider horizontal clearances for locks, narrow channels, and bridges. Drawing your vessel to scale right on the chart is one way to get a sense for the constraints in unfamiliar waters. Having scrutinized the route, the passage planner must appraise the effect of environmental factors such as weather patterns—short-term, long-term, and seasonal. The planner must take into account regulatory constraints. Are the right licenses aboard, and do the load line certificate, stability letter, and COI permit the vessel to operate this way? He or she must consider the practical capabilities of the vessel and its crew. What are the maneuvering characteristics, the steaming range, and the seakeeping qualities? What are the crews' experience levels, and are they fit for duty? Appraisal involves tallying the knowable factors and weighing the reasonable "what ifs" to answer two basic questions:

1. Is the voyage legally and physically feasible?
2. If so, what are the risks and special considerations?

A detailed appraisal enables the planner to make informed decisions about risk, and then to use that information to construct the actual plan. Sometimes an appraisal is done in advance of a hypothetical voyage—could we go there if we wanted to?—but if the voyage is imminent, the appraisal must look at the imminent circumstances.

Many vessels are built, crewed, and certified to ply a specific route in a specific trade, such as towboats on the Western Rivers of the United States or offshore supply vessels running to an oil field. When this is the case, many of the questions above are easily answered. Other vessels may range widely, but make certain transits on a regular basis. If a ship has made the voyage in question before, the appraisal is greatly simplified. It then becomes a matter of researching any changes since the last trip and weighing their significance. Most runs are routine, yet most accidents occur on routine runs.

Some voyages involve going somewhere completely unfamiliar—perhaps a different country, coast, or ocean. These require the most extensive appraisals. All the items mentioned above must be researched but without the benefit of personal experience and local knowledge. On a long voyage, perhaps to a different hemisphere, seasonal differences require a closer look: It may not be hurricane season where you are, but is it where you're going? Personal experience is part of the appraisal. When there is a high level of familiarity with all aspects of the operation, there is reason to believe the voyage can be successfully repeated. But merely having done something before does not constitute a true appraisal. Some sort of fact-finding or deliberation is always warranted, especially where tolerances are small and the consequences of failure are great. Even a familiar voyage requires some scrutiny because things do change.

The Tug *Scandia* and the Barge *North Cape*, 1996

On the evening of January 18, 1996, the tug *Scandia* got underway to make the 170-mile run from Bayonne, New Jersey, to Providence, Rhode Island. The *Scandia* was towing the barge *North Cape*, which was carrying 96,000 barrels of heating oil. Shortly before sailing, the barge's windlass was taken off for repairs. Since the windlass was not required equipment for the uninspected vessel, a jury-rigged system was put in place to hold the 6,000-pound anchor during the passage.

The six-man crew was experienced. The captain had logged more than twenty years on tugs and the mate had seventeen years of experience, much of it on tugs. The engineer held a license for vessels of any horsepower and had thirteen years of experience, including several as chief engineer. The remaining three crew members had between ten and fourteen years in the marine industry.

On the morning of January 18, the National Weather Service forecast a severe and fast-moving winter storm approaching southern New England, including the exposed coast of Rhode Island that lay along the *Scandia*'s route. Approximately 15 hours before the tug's departure, a private weather service provided the captain with a similar outlook: conditions would begin deteriorating on the night of January 18, followed by sustained southerly winds of 35 to 45 knots with higher gusts the following day. Seas would be 4 to 8 feet, but "much higher on waters exposed to the south." By 0250 on January 19, while the vessel was still within the protected waters of Long Island Sound, the Coast Guard issued a revised warning: southerly winds of 45 to 55 knots with higher gusts, and seas 15 to 25 feet. The timing and track of fast-moving storms are notoriously difficult to pinpoint, and therefore call for particular caution. By noontime on January 19, all vessels in the company's fleet from New York to Rhode Island were seeking shelter from the storm, as the whole coast battened down for what was coming. The *Scandia*, lacking a reliable means of anchoring its tow, continued east, beyond the shelter of Long Island Sound, to the open waters of the Atlantic that were predicted to receive the brunt of the storm.

At approximately 1320 on January 19, the *Scandia* and its barge were just 4 miles from Narragansett Bay. The weather was worsening but was not yet hazardous. If the captain was banking on beating the storm, he was nearly there. Then a fire broke out in the engine room.

The chief engineer and the dayman attempted to fight it with extinguishers. Another crew member tried to secure the hatches and doors to the engine room fidley (the upper part of the engine room) but couldn't. The captain thought that the engine room had a fixed CO_2 system but now found out that it didn't, and the main fire pump was not accessible due to the fire. There was a semiportable CO_2 system just inside the fidley, but the engineer was the only person who knew how to operate it and he was unable to enter the space due to smoke and heat. Less than a year earlier the crew had extinguished a major engine room fire while dockside by using turnout gear and breathing apparatus borrowed from the local fire department. After that incident, the captain requested turnout gear and breathing apparatus for the vessel but, like the windlass, the equipment was not required and was not put aboard. Complicating the firefighting effort was the fact that no one but the engineer seemed to have any knowledge of the remote fuel and ventilation shutoffs on deck. In any event, none were activated so the fans continued to supply fresh air to the fire.

Initially, the fire was confined to the fidley, but as the tug lurched and wallowed in the building seas, an unsecured cabinet in the upper engine room toppled over, jamming open a door that connected the engine room to the rest of the vessel. Smoke and heat filled the wheelhouse, forcing all hands out on deck on a winter afternoon. The captain ordered the crew to retrieve their survival suits, and

The route of the *Scandia*, showing the approximate location of its grounding. (National Transportation Safety Board)

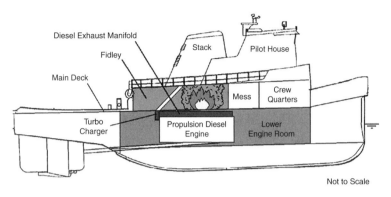

The location of the fidley—the upper part of the engine room where the fire began on the tug *Scandia*.
(National Transportation Safety Board)

shortly before 1400, he reported the situation to the Coast Guard and prepared to abandon ship. Due to flames and smoke, however, the crew were unable to reach the life raft so they gathered on the bow to wait.

At 1509, shortly after a Coast Guard cutter arrived at the scene, an explosion engulfed the entire wheelhouse in flames. The crew leaped into the water, where a rescue swimmer helped them to safety. Winds were now 35 knots and seas were 12 feet from the south. Salvage tugs were en route but were hours away, and in the meantime efforts to get control of the tow failed.

No one anticipated the chain of errors that doomed the *Scandia* and the *North Cape* after they left Bayonne, New Jersey, as a low-pressure system bore down on the East Coast. The result—millions of dollars of damage to the vessels, destruction of a productive fishing habitat, destroyed careers, and the human toll for all involved—was devastating. The failure to appraise the potential effects of shipboard deficiencies as well as severe weather led directly to this incident. (Associated Press)

The abandoned *Scandia*, its engine still in gear, continued on an easterly heading while being steadily driven shoreward.

At 1700, with darkness falling, the mate and the engineer of the *Scandia* were put aboard the barge to deploy the anchor. By this time the winds were 40 to 50 knots and seas were 20 to 30 feet out of south, as the revised forecast had predicted. In the failing light and lacking the necessary tools, the men were unable to release the jury-rigged anchor. An hour later the *Scandia* and the *North Cape* ran aground. Eleven of the *North Cape*'s fourteen tanks ruptured, spilling 828,000 gallons of heating oil onto the beaches of Rhode Island. Conditions soon subsided as the storm rolled onward.

The damage to the two vessels totaled $5.1 million, and the shoreline where the vessels ran aground was devastated. At the time, the State of Rhode Island had a half-billion-dollar fishing industry that employed some 35,000 people. The oil came ashore in the heart of a unique and sensitive string of federally protected estuaries that were breeding grounds for all manner of marine life. Millions of juvenile lobsters, shellfish, and finfish were killed, along with seabirds and other wildlife. In the aftermath, the company pled guilty to felony counts of negligence and agreed to pay $16 million in damages; the president of the company and the captain of the *Scandia* also pled guilty to charges of negligence. The company never recovered, and within a few years it ceased to exist.

Many lessons can be learned from what happened aboard the *Scandia*, only some of which relate to appraisal. Once fire broke out, a cascade of events caused the situation to spin out of control. This was not a case of navigational imprecision or an inherently risky route. But steaming into severe weather along a lee shore with a hazardous cargo and critical equipment unavailable showed a failure to consider the big picture. There was no weighing of the reasonable "what ifs" with an eye to the potential consequences. The episode could have been worse—the lives of the crew and rescuers were spared—and yet it was entirely avoidable.

Failure to Appraise

The storm that caught the *Scandia* was well forecast. Vessels all along the coast were suspending operations on account of it. The intense, fast-moving system warranted careful consideration. When it comes to weather warnings and other unwelcome news, however, there is sometimes a tendency to hope that the reports are exaggerated. This may be especially true on a familiar run. All experienced mariners witness storm systems that don't live up to their billing, and when they have a job to do it is easy to view warnings skeptically. But for every weather system that peters out or veers away, there is another that does the opposite. In this case, the forecast was spot on, and plenty of mariners along the coast had appraised its potential more carefully.

To anticipate that this particular storm would lead directly to a massive oil spill was impossible, but no extraordinary foresight was required to see the potential consequences of the forecast. Be it a fire, a machinery failure, an injury, or a parting tow wire, things go wrong in bad weather that don't go wrong in good weather. If they do go wrong in good weather, they are less likely to have disastrous consequences. More than any other dimension of seafaring, bad weather exposes our weaknesses and poor choices. Foul weather probes and tests the seaworthiness of boats, equipment, and people. When things go wrong in good weather, as they often do, it may be challenging but we are better positioned to manage events. When caught midocean by severe weather there is no choice but to ride it out, but at least there is room to do so. The captain of the *Scandia* approached this transit as if it were a routine trip, when there was reason to believe it wouldn't be.

The absence of a rigorous appraisal was not limited to the weather. The captain also did not attach significance to the lack of a functional anchoring system for the barge. A pre-underway equipment checklist would have prompted greater consideration of this deficiency. Though the *North Cape* was not required to have an anchoring system, it had been given one because it was a good idea; anchors have a history of being useful. Sailing without a working anchoring system was a bad idea, but not so much because the crew couldn't deploy it at the height of the storm—the odds of success at that point were low. More important, the deficiency limited the captain's options at *all* stages of the voyage. The absence of a functioning windlass meant that anchoring for weather, fog, an emergency, or for any other number of reasons was not as viable as it normally would have been. The tow was unseaworthy from the beginning, even without a storm bearing down. A more thorough appraisal would have highlighted this.

The lack of fire training, fire fighting equipment, and general unreadiness for sea aboard the *Scandia* is separate from passage planning, but these issues highlight the manner by which each voyage successfully completed reinforced the false impression that this was a professional operation. In addition to the deficiencies already noted, the *Scandia* carried a clothes drier that ran continuously unless it was manually turned off, and a mysterious electrical short had been giving the crew electrical shocks for months. These facts combine to create the impression of an organization incapable of appraising either the long-term or imminent risks of operating this vessel. Such a culture is bound to result in small margins of safety in all matters, including passage planning. While the emphasis of passage planning customarily falls on navigational precision, this case shows how genuine appraisal must look beyond the obvious concerns such as under-keel clearances.

Planning

The planning stage involves taking the information gathered during the appraisal and building an actual road map for the voyage. The road map includes route fundamentals such as distances, courses, waypoints, physical hazards, and key navigational aids. On a familiar trip we can to reuse waypoints and courses, but weather, tide, current, river stages, draft, visibility, equipment status, and experience levels can change in significant ways. The planner must consider updated information concerning charts, publications, navigational aids, and regulations. A fully developed plan will make use of landmarks, natural ranges, bearings, and other charted features to aid in monitoring progress. With this information the passage planner can begin to form ideas about optimum arrival times at key points and calculate desired speed of advance accordingly. He or she can identify areas of high risk that warrant extra precautions and identify appropriate watch conditions. Once the route is built, a passage plan can expand in many directions, depending on the nature of the voyage. Details concerning pilotage requirements, VHF channels, vessel traffic services, transits of bridges, canals, and locks, and regulations and security measures become part of the plan at this stage.

A passage plan that contains too much detail will swamp watchkeepers in minutiae. But a plan that contains too little detail will fall short as a tool for guiding decisions. Each organization must strike a balance that reflects the practicalities of the specific operation.

The *Safari Spirit*, 2003

The *Safari Spirit* made its living plying the waters of the Inside Passage of the Pacific Northwest. The company specialized in high-end adventure cruising that combined fine dining and comfortable appointments with excursions through some of the most remote and spectacular cruising grounds in the world. At 105 feet, with a draft of just 6½ feet, the *Safari Spirit* was more nearly a passenger-carrying yacht than a cruise ship. It carried a maximum of twelve passengers, which meant that its operators could customize the experience to the preferences of the passengers and take advantage of any special opportunities that came along. The company promoted this aspect of the experience in its literature: "Your itinerary is flexible enough that your captain can change course when word of whale sightings comes over the radio, or slow the engines and linger in a tidal inlet if bears are out foraging on the beach." The *Safari Spirit*'s relatively small size meant that it could go just about anywhere.

Generally, the *Safari Spirit* traveled during the day and anchored at night. This schedule meant the passengers could enjoy the scenery in daylight while allowing rest for the captain, who held the only required license aboard. The captain had twenty-five years of seagoing experience, ten of them in command. The only other person aboard with specific marine duties was a "mate/engineer," who also functioned as a deckhand.

On the morning of May 8, 2003, the *Safari Spirit* was anchored in Kisameet Bay, British Columbia, six days into a fifteen-day trip from Seattle to Juneau, Alaska. The start and finish of the voyage were firm, but the itinerary between these two endpoints was flexible. The captain had come and gone from Kisameet Bay at least four times in the past and was well aware of a submerged rock near the entrance. Not only had he passed it on the way in, he had seen it on other occasions when it was exposed at low tide. At 0705, after running through predeparture equipment tests, the *Safari Spirit* weighed anchor.

The most detailed official chart available for Kisameet Bay is Canadian Chart 3936, Fitz Hugh Sound to Lama Passage, but the scale of that chart is 1:40,000, too small for intimate coastal exploration (small scale = small detail, large area). With such a small-scale chart, local knowledge is needed, and the captain had local knowledge. He compensated for the chart scale by enlarging areas of interest on a photocopying machine and then adding his own details and notes. When combined with local knowledge, this method had proved effective.

For his outbound transit from the anchorage, the captain knew he had to pass clear of the submerged rock at point D (see page 23). To do this he measured

Location of Kisameet Bay on the coast of British Columbia. (National Transportation Safety Board)

Passage Planning 23

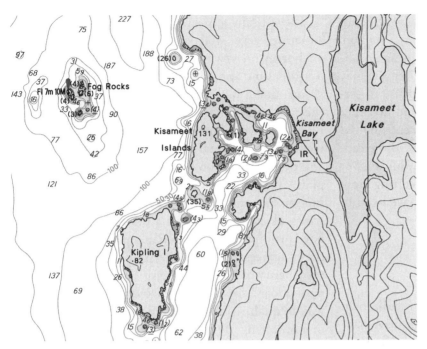

Canadian Chart 3936 shown to scale and greatly enlarged, as was used on the bridge of the *Safari Spirit*. (Government of Canada)

the distance from the south end of the small island (point B) to the south edge of the rock, then added a 120-foot margin of safety, which gave him a total distance of 0.15 nautical mile from point B. He set the variable range marker (VRM) on his radar display for 0.15 nm. When the ring became tangent to point B astern of him, he would be able to make his turn and pass south of the rock. The captain proceeded slowly out of the anchorage using the headland at point E as a visual reference. When the VRM became tangent to point B, he altered course to starboard as planned. Seconds later the vessel rode up onto the submerged rock. The hull was intact, but the *Safari Spirit* was hard aground on a remote isolated rock with a falling 10-foot tide.

The captain evacuated the passengers to shore and set a kedge anchor to steady the vessel on the rock so that it could refloat when the tide turned. But as the tide ebbed and the stern settled, the hull adopted an increasingly precarious angle. Water flooded over the transom and filled the after part of the vessel, causing it to trim still farther by the stern. Three hours after grounding, the *Safari Spirit* slid off the rock and sank in about 70 feet of water, with the stern resting on the bottom and the bow bobbing at the surface. The morning was calm and sunny, with no traffic and excellent visibility.

A depiction of the *Safari Spirit*'s intended track and the actual track that led to its grounding in Kisameet Bay. (A) shows where *Safari Spirit* was anchored. (See the text for an explanation of the other points used by the captain.) (National Transportation Safety Board)

A more detailed passage plan built around multiple references would have exposed the captain's miscalculation and made it easier for him to monitor the route to open water. (National Transportation Safety Board and Shearwood Marine)

The *Safari Spirit* was declared a total constructive loss. Costs related to the incident ran to $3.2 million. After the fact, the captain conceded that he had probably set the VRM inaccurately. Alternatively, he may have mismeasured the distance on the chart, or perhaps he did both. Since the plan contained no other bearings, ranges, references, or measurements, he did not catch his error. When investigators asked the captain what he could have done differently, his answer was simple: he should have checked his plotting more carefully. Another good answer would have been to have a more detailed plan.

Failure to Plan

Given that this incident was due to simple human error, it is perhaps fair to say that it could have happened to anyone. Indeed, it was neither the first nor the last such incident. But it is also the sort of mishap that shouldn't happen to anyone, especially a professional mariner, because it is so easily avoided. Some might conclude that the nature of the *Safari Spirit*'s activities made such an incident inevitable. The flexible and spontaneous nature of the itinerary meant that planning had to be conducted and modified on short notice, which potentially left the captain more vulnerable to navigational errors. Also, the scale of the chart was too small for the sort of cruising this vessel specialized in. These factors are worth considering and could be remedied through better operational procedures, but to condemn the entire enterprise based on them misses the real point.

The problem was not with adventure cruising in remote places, or even with the chart. Such activities are carried out all the time all around the world, usually without incident. The real problem was the sparseness of the planning effort for making the short transit from the anchorage to open water. Straightforward as the transit was in clear, settled weather, the plan was a one-legged stool that was particularly vulnerable to a single-point error. When the captain miscalculated the clearing distance he had no cross-references to expose his mistake. The captain himself was at risk of a single-point error because he was the only navigation officer aboard. That he would one day make a critical error of calculation was predictable—and preventable through a more committed approach to passage planning.

The captain appears to have not appraised the situation very well either—it was just another anchorage on a two-week trip. Visiting the cove was both feasible and reasonable, not a reckless proposition. But a better appraisal would have revealed the need for a better plan. First, the location was remote. Anyone operating in remote places should view all navigational errors as final and irrevocable, and this attitude should be reflected in the way they plan. Second, the bottom was not a gradual sandy incline or a mudflat. It was rock, and the rock he was seeking to avoid was a pinnacle, which should have prompted a more reliable plan. Third, he was the only person aboard licensed to operate the vessel. There was no bridge team. This situation is common aboard smaller vessels, but it meant that there was no qualified second opinion and no one to assist in monitoring the vessel's position on the way out. It was all on one person, which also should affect the way you plan. Fourth, he was departing as the tide was ebbing, and the tidal range was significant. If you go aground on a falling tide, the situation is likely to get worse before it gets better, if it ever does get better. This too should have affected the plan. Last, he was working with a modified chart to overcome a problem of scale. This was not insurmountable but it called for greater care, and this should have affected the plan. Unlike vessels that live their lives plying established commercial routes with buoyed channels, lit ranges, dayshapes, racons, vessel traffic services, and pilots, these waters demand a lot of a navigator. When you add it all up, one can scarcely think of a situation in which a carefully crafted passage plan would be of greater value, even for a short leg.

There are many tools and tricks for planning a tight transit and we will review these in Chapter 4. Most of them were available to the *Safari Spirit*'s captain. The ones chosen on any particular day depend upon the situation. Though there were no navigational aids in Kisameet Bay, clear weather combined with a variety of distinctive coastal features meant visual bearings were readily available. The area also provided excellent radar imaging for setting more than one VRM in conjunction with EBLs (electronic bearing lines). By using any of these to confirm his single VRM, the captain could have improved his chances of catching his error. Though the *Safari Spirit*'s radar display lacked a dedicated parallel index feature, it did have the ability to offset an EBL as a temporary parallel index line.

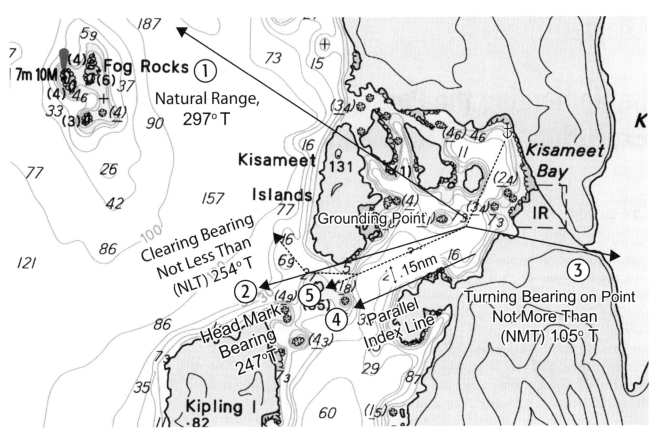

This annotated chart depicts some of the options that were available to the captain of the *Safari Spirit* as a means to double-check his position. These could have been established in the planning stage and used in conjunction with the variable range marker (VRM), to monitor the vessel's position along the track. (The dashed line shows the captain's intended track. Soundings in meters.)

① Natural Range, 297°T on the two islands northwest of the waypoint provide a line of position to indicate the approach of the waypoint.
② Clearing Bearing Not Less Than (NLT) 254°T on the south point of Kisameet Island establishes a margin of safety for monitoring proximity to the submerged rock.
③ Turning Bearing Not More Than (NMT) 105°T on this conspicuous point of land to the southeast of the waypoint indicates the turning point.
④ Parallel Index Line(s): One of these have been shown on the chart, but any number of them could have been set up on other headlands or other visual points to determine whether *Safari Spirit* was left or right of track after making the turn.
⑤ Head Mark Bearing 247°T.

This, perhaps more than any other tool, would have enabled him to see the error in his initial plan while providing a better guide for making his turn.

Finally, an excellent technique when operating in tight quarters with small-scale charts is to launch a small boat and send it ahead with a lead line. This practice has been around for centuries; over 200 years earlier a young William Bligh performed exactly this service along the same coast for Captain James Cook with no charts at all. Soundings can be taken and (nowadays) radioed back to the ship until well clear of the hazard. Such measures were not absolutely necessary in this case, but some better combination of tactics was.

Except for the limitations imposed by the chart scale, which the captain had remedied, the sinking of the *Safari Spirit* is entirely attributable to human factors. All other factors favored a safe passage. This was a case of the planning being too casual and too limited. The uncritical application of incorrect information was magnified by the captain's reliance on a single reference point. Good tools stayed in the box, and the captain grounded on a rock that he had actually made a conscious effort to avoid. Just a few more minutes spent on planning could have avoided this incident, and over $3 million in damages.

Chapter Three

Implementing the Passage Plan: Execution, Conferring, and Monitoring

Execution

Though a passage plan is never set in stone, some aspects of it cannot be finalized until a departure window has been confirmed. The *execution stage* of passage planning occurs shortly before departure, when you can decide timing and tactics with greater accuracy. For instance, you cannot establish ETAs until the departure time is known. You may have information on the tidal ranges and the current patterns affecting the voyage, but predictions for the precise time and date of the coming transit must wait until the time and date are known. Though you will have considered the roles of daylight, weather, and visibility during the appraisal and planning stages, their impact remains hypothetical until you know when the ship will be getting underway. You should also consider traffic conditions at this point, especially around navigational focal points. Sometimes these variables are critical enough to actually drive the departure time. If a transit is tide dependent, or is best accomplished at slack water, or with a fair current, or stemming the current, or in daylight, or with the sun at your back, or with assistance from tugs and pilots, then vessel movements will need to reflect those realities. Execution includes once more ascertaining the reliability and condition of your resources: equipment (especially navigational equipment), people, and information. The execution stage is the time for refining details and making final decisions.

The Towboat *Anne Holly*, 1998

Operating on the Western Rivers of the United States is different from operating elsewhere. Rather than making distinct voyages across blue water from one port to another, towboats operate more like long-haul trucks, loading and unloading continuously along a riverine highway, dropping off and picking up barges. Crews change out along the way, turning the vessel over like a baton in a year-round relay race. Navigation on the rivers presents a ceaseless procession of challenges with little of the relative relaxation afforded by open water. The intimate familiarity required to negotiate narrow bridges and channels, shifting shoals, complex currents and changing water levels is one reason that watch officers on towboats are commonly known as *pilots*. With a two-watch system, the pilot runs the boat for half of the time with little or no supervision from the captain; therefore, he or she must possess as much skill and knowledge as a captain. The tows are long, often exceeding the lengths of the largest oceangoing ships. The captains, pilots, and crews have evolved special techniques for coping with the challenges of keeping commerce moving under conditions that would undo the most seasoned bluewater mariner. For all these contrasts with saltier realms, human error and the underlying principles of BRM play as large a role here as anywhere else, they just look different.

On the evening of April 4, 1998, the Mississippi River at St. Louis was running at 6 knots. Due to the spring thaw and heavy rains hundreds of miles to the north, the river gauge at St. Louis read 31.5 feet, 18 inches above floodwater conditions. Because of this, the Coast Guard had established a safety zone for the port of St. Louis that restricted southbound (downriver) tows over 600 feet to daylight transits only. The towboat *Anne Holly*, with 5,600 horsepower, was preparing to take a fourteen-barge tow to St. Paul, Minnesota, over 600 miles to the north. The total dimensions of the tow were 1,149 feet long, including the 154-foot towboat, by 105 feet wide. On this particular night the steering light—a small, blue light typically mounted on the centerline at the head of the tow to help monitor heading changes—was out.

The captain of the *Anne Holly* had nearly forty years of marine experience, most of it on towboats,

The towboat *Anne Holly* underway on the Mississippi without barges. (Richard E. Dunbar)

The view from the bridge of the *Anne Holly*; it's a long way to the end of the tow. (National Transportation Safety Board)

and had been licensed for twenty-five of them. He had transited St. Louis many times under various conditions, including high water, though less frequently under the particular confluence of circumstances prevailing on this occasion (night, floodwater conditions, and a large tow). The pilot running the other watch had been working on the rivers for twenty-four years and had been licensed for eighteen of them. Both the captain and the pilot were aware of the high-water conditions.

At St. Louis four bridges span the Mississippi River in a 1.2-mile stretch. One of these, the Eads Bridge, is considered one of the most difficult spots

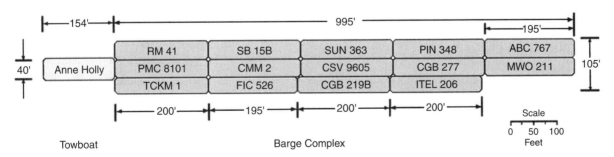

Schematic of the 154-foot towboat *Anne Holly* and the 995-foot barge complex showing the fourteen barges pushed up the river at the time of the accident. (National Transportation Safety Board)

on the Western Rivers because the structure curves downward from the centerline to the bridge piers, restricting the amount of vertical clearance on either side of the centerline, especially when the water is high. The captain called it "the worst bridge on the Mississippi River. Everybody hates it." The pilot described it as "hazardous." But the high water and the challenges posed by the bridges did not affect the captain's intention to get underway, nor did the two men discuss whether or how the conditions might impact their plans. A reality of the two-watch system is that there is little opportunity to confer except at the change of the watch. The two watches tend to operate independently of each other. It was not customary to consult shoreside management on such matters and there were no procedures to guide the captain's thinking. The Coast Guard had a restriction on downbound nighttime transits at the time, but the *Anne Holly* was headed upriver so it did not apply, though it did provide a clue as to the prevailing conditions. The captain thought it would be a good idea to have an assist boat—a smaller towboat used to help maneuver a large tow in difficult situations—but this was more of a private notion than part of any structured plan. Without confirming the availability of an assist boat he cast off when the tow was lashed up.

The *Anne Holly* got underway at 1830 and began heading upriver. Soon after, the captain radioed the fleet dispatcher and requested one or two assist vessels to help maneuver the tow through the downtown bridges. This was a common practice, and it was prudent under the circumstances. It was at this time that he learned there were no assist boats available. Since the trip had already begun, he decided to keep going. The night was clear, calm, and all systems were functioning normally.

The *Anne Holly* transited the first two bridges without difficulty. The restricted clearance of the Eads Bridge meant the towboat would have to stay very close to the center of the span when passing under it. Another challenge of the Eads Bridge is that it is only 1,200 feet downriver from the next bridge. With a tow of 1,149 feet, the captain had to maneuver for the next bridge while still beneath the Eads Bridge. The captain's task depended essentially on visual cues and a feel for the tow; electronics were superfluous. But, as we know, visual cues, and therefore feel, are degraded by darkness. Backscatter and glare from city lights interfere with depth perception and add to the challenge. The captain used spotlights to pick out the dayshapes marking the bridges, but with the rapid current and erratic eddies, an assist boat or two would have been useful at this point.

The location of the *Anne Holly* accident on the upper Mississippi River in St. Louis. It had cleared the first two bridges and was heading for the third, the Eads Bridge, which has an understructure that requires the operator to stay very close to the center span of the bridge, especially at times of high water. (National Transportation Safety Board)

As the lead barges passed under the Eads Bridge, the captain maneuvered to keep the towboat aligned with the center of the span while setting up

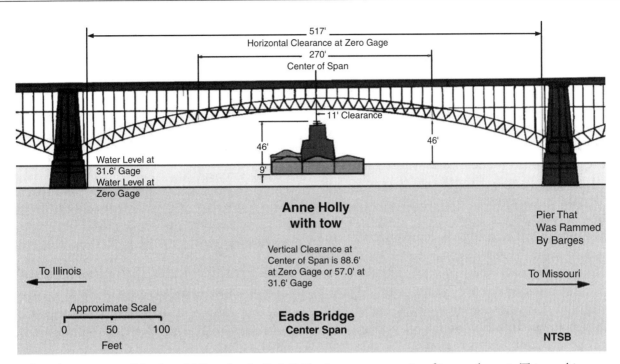

This schematic view of the *Anne Holly* under the Eads Bridge is not representative of its actual transit. This graphic was used in the National Transportation Safety Board's report and was intended to convey the relative heights of the *Anne Holly* and the bridge as well as the water depth at the time of the accident. (National Transportation Safety Board)

for the next bridge. A steering light would have been helpful, but it wasn't lit. Everything was going fine but currents are often erratic around structures, and, suddenly, the tow sheered to port and slapped down onto the western bridge pier, snapping wires and scattering barges. Later the captain suggested that this was caused by a *pop rise*, a sudden and unpredictable rise of the river, but sensors and gauges in the area showed no pop rise that night.

When things go wrong, they go wrong fast. Just downriver of the Eads Bridge, on the St. Louis side of the river, was the 380-foot *Admiral*, a permanently moored casino vessel, with some 2,200 patrons and staff aboard on this Saturday night. Propelled by the 6-knot current, three of the *Anne Holly*'s runaway barges smashed into the *Admiral*. One by one, the *Admiral*'s mooring wires parted under the impact and, as each one gave way, the strain on the remaining wires grew. With little to hold the upriver end to the bank, the *Admiral* swung out into the river, tearing away the three gangways that provided the only escape routes, severing telephone, gas, and electrical connections, and blacking out the ship. The captain of the *Anne Holly* responded swiftly, issuing a Mayday call, disengaging his remaining barges, and maneuvering to pin the *Admiral* against the riverbank, where it came to rest 180 degrees from its original position and hanging on a single wire. Luckily the *Admiral* did not tear away and drift downriver toward the other bridges. There was no gas explosion or electrocution. Since the *Anne Holly*'s barges carried no hazardous cargo, none was released. Everyone was evacuated safely. Sixteen people were hospitalized, but no one was killed. Still, $11 million of damage is not the kind of luck that advances a career or keeps a company out of the papers.

The *Admiral* was pinned against the riverbank by the *Anne Holly*. There was $11 million in damage. (National Transportation Safety Board)

Failure to Execute and Confer

A number of factors led to the *Anne Holly* incident, including a strikingly unconcerned approach to appraisal. Indeed, investigators concluded that the overriding cause of the accident was the captain's poor judgment and poor decision-making in attempting the transit with a large tow, in darkness, under high current and flood conditions, and without an assist boat. Change any one of those factors, and it's just another day on the river.

How do you defend against poor judgment and decision making? The answer is complex but it usually starts with having adequate experience. The captain had adequate experience, but that alone was insufficient. Another defense against poor judgment and decision making is having the right information at the right time. This is a function of passage planning in general and, in this case, the execution stage in particular. Though the captain had most of the information he needed due to his personal experience, he found himself facing the Eads Bridge without the assist boats he wanted. A more deliberate approach to execution would have informed him of the unavailability of assistance while still at the dock and afforded the chance to weigh the implications.

Conferring with the pilot, who was also very experienced, was not a customary practice for this kind of decision. This was partly due to the watch rotation but also to a professional outlook that did not prioritize resource management in this way. Yet a conversation with the knowledgeable pilot, who also had a stake in the outcome, might have produced a different view of getting underway with so many unfavorable factors. Once it was clear that no assistance was available, a nighttime transit through the city under the Eads Bridge on a flooding river might have looked different to one or both of the licensed deck officers aboard. The lack of a steering light could have been addressed and corrected or added to the reasons to reconsider the plan. There were at least two good alternatives: wait until an assist boat was available or wait until daybreak. But it is difficult to give alternatives proper consideration once things are set in motion, partly due to momentum and partly because people become committed to their plans, including the bad ones. No one wants to appear indecisive. The lack of conferring aboard the *Anne Holly* meant there was one less mechanism to prompt the captain to reevaluate what turned out to be the most important—and worst—decision of his career. He made a personal decision to carry on and hope for the best, but in doing so he put the careers and lives of many others at risk, however unintentionally. We will never know if a discussion with the pilot or anyone else would have resulted in a different outcome, but we do know that the captain acted entirely on his own and he made the wrong call. The watchkeeping structure was not ideal for the kind of planning conference that is possible elsewhere, but that does not diminish the need to consult and confer—in other words, manage your resources.

Conferring

Conferring is the act of sharing information and openly discussing ideas and concerns. Conferring can, and should, happen at all stages of passagemaking, but it is particularly appropriate at the execution stage. Conferring is central to creating the shared mental model.

Conferring before getting underway, particularly for a routine run, may seem like a needless step when there is so much business to take care of. And if everyone knows what they are doing, what is there to talk about? But most serious accidents occur on routine runs and involve people who very much know what they are doing. Considering the role of normal human error, talking over the plan is not necessarily a poor use of time for people of all skill levels. It is precisely when people are hustling to get the job done that misunderstandings, complacency, and a blurring of priorities are likely to occur.

Conferring is the process of making the plan transparent to those who will be executing it. Like SOPs and the passage plan itself, the *planning conference* is a communication device intended to help catch mistakes and prevent misunderstandings. Good planning usually yields good results, so it is only logical to have a mechanism for making sure that the plan is clear to all who will carry it out. A system built on assumptions is prone to failure, and we have already seen how written words can be misconstrued.

Conferring is nothing new. It has always taken place between crew members with varying degrees of formality. Much of it happens without prompting because it is so obviously necessary to get anything done. The point of focusing on it here is to draw attention to the importance of these exchanges and to make them a conscious part of business as usual.

Operating large tows on riverine highways like this leaves little margin for error. In this case, execution involves taking into account the effects of current, river stage, fog, and other waterborne activities, especially around choke points. (U.S. Army Corps of Engineers)

Conferring is the act of pooling knowledge and experience. It is the practical acknowledgment that two (or more) heads are better than one and that everyone makes mistakes. In this sense, conferring provides another line of defense against the single-point error.

Whether it takes the form of an organized meeting or a brief, face-to-face exchange, conferring has both obvious and subtle benefits. It can help alleviate the isolation of a decision-maker in a high-risk situation, such as that faced by the captain of the *Anne Holly*, by reminding that person of resources and alternatives that might otherwise be overlooked. It provides a second opinion that may reveal an oversight, a fundamental flaw in the plan, or simply shed more light on the inherent risks. Sometimes your gut tells you that you are pointed toward trouble, but momentum and time constraints cause you to ignore your own misgivings. Conferring provides an opportunity for others to raise valid concerns or share new information, or simply to confirm what your gut is telling you. Given the opportunity, some people will be more inclined to speak up than others, but if no opportunity is created, nothing will be discussed.

Conferring can have a professionalizing effect as it helps solidify the notion that everyone owns a share of the outcome, good or bad, and that any kind of incident is bad for everyone, not just the person on watch when it happens. The act of putting a plan forward forces us to think it through more carefully than we might otherwise, and it tends to pour cold water on private temptations to wing it. The process of preventing errors starts with the plan itself, but it plays out again when the plan is shared with other people whose licenses are also on the line.

Obstacles to Conferring

Conferring is the hardest stage of passage planning to pull off, primarily because there never seems to be enough time, especially with a two-watch rotation. Sometimes conferring is difficult to attain for other reasons. On the *Safari Spirit* (see Chapter 2), the captain was the sole operator aboard, which put him at a distinct disadvantage as far as conferring goes. This was all the more reason to get everything else right.

Another reason conferring sometimes doesn't happen is that not everyone likes the idea. Some people just want to be told what to do, and many other people are not interested in a second opinion. This is especially true when an opinion contradicts your own, or when it comes from someone you do not have a positive professional rapport with. Yet it may be precisely these situations, when the shared mental model is most elusive, that conferring has the greatest value. When a lack of conferring marginalizes good resources, we should not be surprised if things do not go well.

Another reason people fail to make time to confer is that it may seem more dispensable than other precautions: If everyone knows what they're doing, what's there to talk about? While there are surely occasions when this outlook is true, the elements of passage planning work together like the mesh of a safety net. If you eliminate certain strands you will still catch some mistakes, but not as many as when the mesh is complete. Like other upgrades to the way mariners do business, conferring must be made a priority for it to yield benefits. When work pressures make conferring inconvenient, consider the potential for it to prevent something bad happening.

The Planning Conference

A planning conference is akin to a safety briefing. It should be face to face, not conducted piecemeal or on the fly. In a perfect world, all crew involved should hear the same information at the same time. That way, if questions or concerns arise, everyone will hear them and have a chance to respond. Since one function of conferring is to prevent single-point errors, participation entails critical thinking. If participants merely rubber-stamp what they hear, the shared mental model will not work as intended, and the operation will be denied the full benefit of all resources.

Anyone who bears responsibility for executing a portion of the plan should participate in the planning conference. Someone who is in the loop can almost always provide better support than someone who isn't. If a machinery problem were to render the vessel especially vulnerable at a certain point in the trip, then it makes sense to involve the engineer so that critical systems can be at maximum readiness, and the engineer will be poised to respond if anything goes wrong.

The information and the level of detail covered in a planning conference vary with the situation. For a routine trip involving the regular crew and no special circumstances, a short discussion between key players over a cup of coffee may suffice, but a multivessel anchor-handling job should involve considerably more detail. A passage planning conference addresses the plan itself: the route, areas of high risk, weather, traffic, etc. This is the time to clarify where and when certain things must happen and who will do them. A review of the readiness of the vessel, crew, equipment, and the progress of any special preparations are also part of developing the shared mental model.

Conferring has a place in any aspect of vessel operations that requires heightened awareness and coordination. The amount of time invested in conferring is a question of the potential returns. Inevitably, people will arrive at different answers. Bunkering, tank entry, loading and discharging, and even maintenance are examples of routine activities that benefit from conferring. For people to function as members of a team, they have to be treated as such. Conferring reminds individual crew members of the importance of each person aboard. It can help you get the best from your people, and the smaller the crew, the more each person counts.

Monitoring

Monitoring is what you do on watch. It involves checking your position, talking on the radio, and analyzing traffic. It involves taking the measure of environmental changes (tide, current, weather, visibility, etc.) and making adjustments. Monitoring involves making decisions, which may include calling the captain. Monitoring is also maintaining the logbook and following the passage plan. Monitoring involves gathering information about the near future so you can pass on the best possible information to the next watchkeeper.

Monitoring is when passage planning earns its keep. It is the act of putting the passage plan into service and using it to stay on track. Remember, it is infinitely easier to spot what is *not* supposed to happen if you know what *is* supposed to happen. We are all vulnerable to inadequate monitoring at times. Situational awareness wanes and then, seemingly out of the blue, all hell breaks loose.

There is no one strategy for effective monitoring, but it always requires being alert. Good monitoring involves balancing competing demands while having some capacity to zoom in on the tiniest detail, recognize its significance, and then zoom back out to reassess the big picture. To illustrate, we'll look at two very different incidents in different parts of the world. As in all such incidents, we can point to multiple BRM issues, but ineffective monitoring is prominent in both.

The Tug *Mercury* and the Barge *101*, 1994

On December 30, 1994, the tug *Mercury* was towing the Barge *101*, with 62,404 barrels of diesel oil aboard, through Puget Sound from Vancouver, British Columbia, to Jack Island, Washington. Though darkness had fallen, conditions were good when the captain took the watch at 1915. The barge was about 1,000 feet astern and towing well. The captain had more than sixteen years of experience towing in Puget Sound and was familiar with the route. Current tables showed that the tidal current would be setting south between 0.5 and 1.5 knots in the Strait of Georgia at the time of the *Mercury*'s transit. The *Mercury* was heading southeast, making 8.5 to 9.0 knots. A quick mental calculation shows that the current could potentially set the tow as much as a mile south of its intended track over the course of 1 hour if not compensated for. Of course, it could be less, but it could also be more. This much was foreseeable.

At 2109, just north of Patos Island, the captain called VTS and learned that there was no large commercial shipping in the area for the time being. Basically, he had that part of the Puget Sound to himself. The southbound traffic lane was over a mile wide, gradually tapering to half a mile in the vicinity of Clements Reef. Despite having ample room to maneuver, the captain skirted along the southern edge of the traffic lane. The company's written instructions required that fixes be plotted at a minimum of every 30 minutes in these waters, but the captain instead chose to navigate by eye, occasionally monitoring visual and radar bearings but not plotting them. The chart he was using was 1:25,000, not the

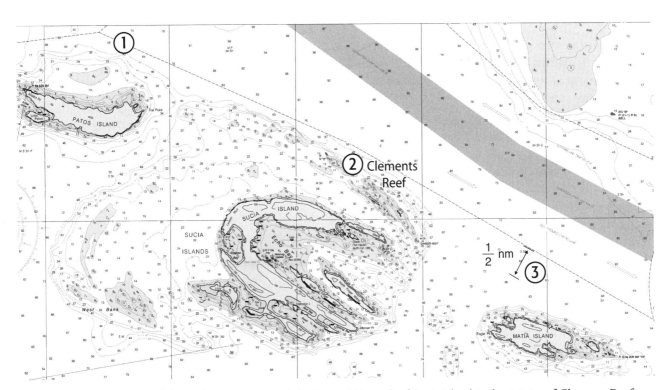

Approximate locations of the towboat *Mercury* and the Barge *101* north of Sucia Island in the vicinity of Clements Reef. (NOAA Chart 18431, 1:25,000; soundings are in fathoms.)

① Approximate position of *Mercury* at 2109 VTS check-in
② Probable site of grounding
③ Approximate position of *Mercury* when hailed by VTS

largest scale available, and though it had marks on it from previous trips, he had not drawn a track line for this trip. The wheelhouse was equipped with a functional GPS receiver as well as a depth-sounder, but no fixes were plotted and depths were not monitored. Given the captain's familiarity with the waters, this was an easy trip.

While passing Clements Reef, a marked hazard, the captain made two small course changes to port to counteract the southerly set of the current and to stay clear of the shoal water to starboard. He later said that the course changes were approximately 5 to 7 degrees each, but that cannot be confirmed because these, too, were done by eye and were not logged. The captain believed he had cleared Clements Reef by about a quarter of a mile, but given his navigational methods his actual clearance could have been quite different.

About the time the *Mercury* reached Matia Island (see chart), the VTS observer noticed that the tow was about half a mile outside the south boundary of the traffic lane. When the VTS observer radioed to point this out, the captain of the *Mercury* replied, "Roger. Out." The tug immediately altered course to port and reentered the southbound traffic lane.

At about 2345, when the vessel tied up at its destination, oil was seen in the water all around the barge. Subsequent investigation showed that the bottom of the barge was torn and buckled for about half its length on the starboard side. About 40% of the cargo, some 27,000 gallons of oil, was gone.

Because the damage was not detected at the time it occurred, and because the captain had not plotted any fixes, the precise location of the grounding could not be confirmed, but investigators concluded that the barge most likely struck Clements Reef without the captain knowing and was then towed for another 2 hours through Puget Sound.

It does not require hindsight to see that this incident was brought about by a watchkeeper who didn't keep watch. That he had standardized procedures prescribing the minimum frequency of position fixing made it all the more perplexing. Though more advanced electronics might have made the effects of the current more plain, even without such instruments, his task was not difficult. The tools available and his experience were more than adequate to complete the transit safely. Yes, there was some current to cope with, but few would describe it as strong. Vessels that operate at slower speeds, such as the tug and tow, are more susceptible to the effects of current than faster ones, but with sixteen years' experience, this was not news to the captain. Also, the captain would have understood that a tow doesn't necessarily follow in the wake of its tug, and that the tug could safely go places that the loaded barge couldn't. He had plenty of sea room to work with, but because he was not monitoring effectively, he did not make use of it.

The *Maria Asumpta*, 1995

The *Maria Asumpta* was an extraordinary vessel by any measure. Built in Spain in 1858, it had carried cargo and people for 137 years, first under sail, then under power, and then under sail again. The vessel made numerous transatlantic crossings and sailed extensively in the Mediterranean Sea over a remarkably long life. Having been saved from the scrap yard in 1980, it was restored as a brig (two masts, both square-rigged). The brig operated out of Great Britain as a sail-training vessel, where novices and aspiring mariners could learn the ways of the sea aboard a truly authentic tall ship. The captain was the principal owner, and had overseen the restoration, maintenance, and operations of the *Maria Asumpta* throughout its latest career as a sail-training vessel. He had been a naval officer and had since sailed tall ships far and wide for more than twenty years. He knew the coast and his vessel intimately. There was probably no one in the world better qualified to operate this particular vessel.

On the morning of May 30, 1995, the *Maria Asumpta* departed Swansea, Wales, with fourteen people aboard, having disembarked twenty-four

The *Maria Asumpta*, built in 1858, was originally a cargo carrier, but was rescued from the scrap yard and started a new life as a sail-training vessel in the 1980s.
(Emmanuel Le Clercq)

Implementing the Passage Plan 35

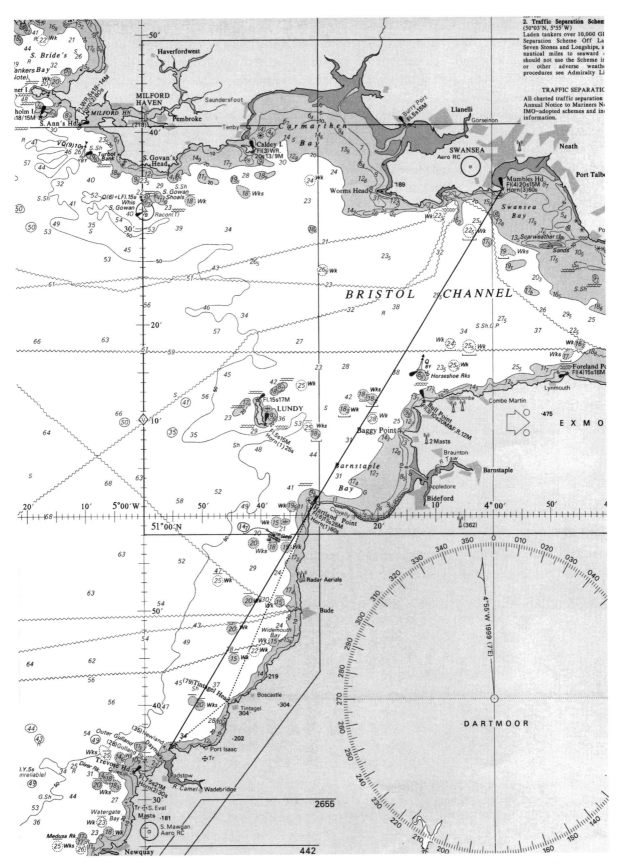

Original track (—) and final track (·······) for the *Maria Asumpta*'s run down the Bristol Channel.
(British Admiralty Chart, with track information from MAIB, *Summary Report*)

students the day before. The vessel was due to arrive at the port of Padstow, Cornwall, in the afternoon. The trip started under engine power but once the wind became favorable the captain set sail and shut down the engines. With the wind northwest, the *Maria Asumpta* proceeded across the Bristol Channel and down the coast of Cornwall on a starboard tack.

The captain laid a course to a point approximately a mile and half off Hartland Point (see chart) on the south coast of the Bristol Channel, and another into Padstow Bay that passed inside Newland, a small island seaward of Padstow. This leg passed west of Roscarrock, a submerged, unmarked rock (see chart). The route inside Newland was not a recognized approach in the *Coast Pilot*, but as an operator of a small vessel, the captain was accustomed to smaller margins of safety. Shortly after passing Hartland Point, however, the captain changed his plan and laid a new course much closer to the lee shore and deeper into the embayed area south of Hartland Point toward Tintagel. Because a square-rigger can neither sail close to the wind nor short-tack to windward, it is vulnerable when embayed on a lee shore. Having a flat-bottomed cargo hull, the *Maria Asumpta* could not perform well to windward compared to other sailing vessels. The captain later explained that he entered the bay because he wanted to avoid the full effect of an incoming tidal stream, and he also wanted his passengers to enjoy the attractive coastal scenery. He could always use his engines to push offshore if need be.

As the vessel neared the coast, the captain plotted positions every 15 minutes. The 1500 position showed that the vessel was inshore of the new track line, which was already inshore of the original track line. The helmsman was ordered to steer closer to the wind in an attempt to take the vessel more out to sea. The captain shifted to a larger-scale chart and plotted two more positions, about 8 minutes apart. These showed the vessel to be still inshore of the second track and setting up to pass far closer to the Mouls (see chart), a rock islet, than intended. Maintaining the original track would now require a large course change that

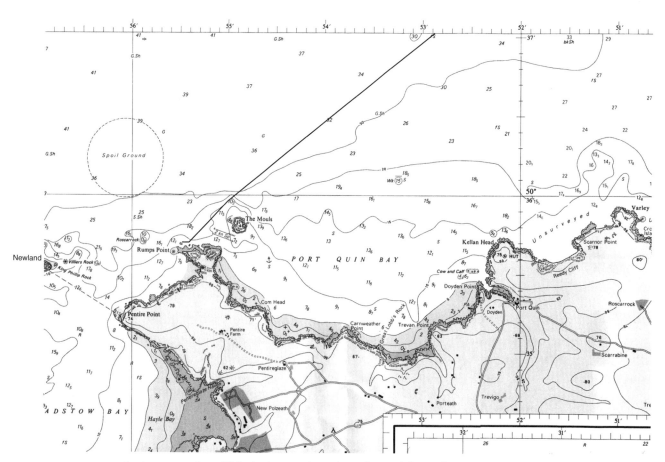

An interpretation of *Maria Asumpta*'s track made good down the Bristol Channel. (British Admiralty Chart, with track information from MAIB, *Summary Report*)

was not possible with the prevailing wind direction and the flooding tide. Under sail alone, the original plan was now out of the question. At one point a crew member suggested tacking out to sea to get clear of the shore. Since the vessel was running ahead of schedule, there was time to do this. The captain appears to have considered all these factors, but instead of tacking out to sea or firing up the engines, he decided to sail through the slot between Roscarrock and the mainland at Rumps Point.

The Padstow harbormaster was expecting the *Maria Asumpta* and was observing the vessel's progress from shore. Concerned about the captain's approach, he called the ship by cell phone and advised against it. The improvised approach was unnecessarily risky and it contradicted the *Coast Pilot*. Instead, the captain continued to make adjustments to the sails to try to squeeze slightly better performance from the vessel. Eventually, seeing that both wind and the incoming tidal stream were conspiring to set the vessel ashore, the captain ordered the engines started.

Rather than steering back out to sea under power, however, he continued to tweak the sails and make minor heading changes in an effort to get around Rumps Point, using the engines to mitigate leeway. Five minutes later, the engines quit and could not be restarted. The captain said to the helmsman, "This could be serious." It was. The *Maria Asumpta* had been sailed into a corner that it could not be sailed out of. The wind was light, the current was setting it on shore, and a moderate groundswell interfered with progress. There was no margin of safety at all. The venerable sailing ship piled onto the rocks at the foot of Rumps Point while onlookers atop the cliff, who had gathered to greet the ship, stood aghast. The old wooden ship was slammed repeatedly by the swell as the rigging crashed down and the hull opened up. Within 5 minutes, the *Maria Asumpta* ceased to be a ship. Passengers and crew leaped into the water and were battered by the wreckage. Some clambered onto the rocks only to be dragged back in by the surf. The disaster unfolded in broad daylight, in clear conditions, and with time to spare. Three crew members drowned. The captain was later sentenced to eighteen months in prison.

Prison may sound like harsh punishment for what could be viewed as a mechanical breakdown. Mariners rely on engines to ensure safe navigation, and, sooner or later, they all experience engine failure. If engine failure occurs in the wrong place, the ship can

The *Maria Asumpta* after it came to grief, with the Mouls shown in the background. (PPL)

be lost. It's really that simple. This engine failure happened in a place where the ship had no business being, a place the ship had arrived at due to a misguided approach to passage monitoring. The pilot who was waiting to guide the *Maria Asumpta* into Padstow later remarked, "When I realized where she was, I was dumbfounded." Another witness commented, "It mystifies me what she was doing there." The captain had violated a golden rule of sailing: avoid the lee shore. When the engines died, the full benefit of good seamanship was no longer available to him. The judge later said to him, "You betrayed [the victims'] trust by showing contempt for the very dangers they trusted you to avoid."

Failure to Monitor

The captains of the *Mercury* and the *Maria Asumpta* found trouble in very different ways. The *Mercury*'s captain simply wasn't doing his job at a basic level. Despite ample sea room, clear instructions on position fixing, and with adequate means, he made no sustained attempt to fulfill the most elementary responsibilities of a watchkeeper. VTS appeared more alert to the *Mercury*'s position than the captain, who was chided back into the traffic lane. His resources—GPS, radar, charts, and buoys—were barely tapped. The theoretical cure for such a state of affairs is relatively simple: do your job, all parts of it, and especially the part about knowing where you are. Experience was another valuable resource, but in this case it took the form of complacency, that most human response to the familiar. It is a lot easier to do the job of monitoring when there is a detailed plan to follow, and this too was lacking.

The situation aboard the *Maria Asumpta* was more complex. The original passage plan was rudimentary but conceivably adequate had it been followed. The captain cannot be faulted for being lackadaisical underway. On the contrary, he was very active. He made numerous decisions throughout the day related to sailing, navigation, wind, current, and the schedule. He plotted courses, changed chart scales, and, as he approached shore, plotted fixes at ever-shorter intervals, just as you would expect from a professional. He monitored progress closely and was deeply involved in the details of directing the vessel—perhaps too much so. Somehow, despite all that activity, he failed to see how he was placing his ship at risk unnecessarily.

From a monitoring standpoint, the captain's big mistake was departing from the plan and not recognizing the significance of it. Doing so initiated a dalliance with a lee shore over which he ultimately lost control. He was monitoring all right, but into danger, not away from it. In a classic case of what is sometimes known as *coning of attention*, the big picture seems to have slipped away from him despite the warnings of others. The captain's fixation with sailing into Padstow, even before the engines quit, distorted his view of the risks. The wind and current conditions did not cooperate, but so subtle were the effects, and so determined was the captain, that he was lured into persisting. Soon he had convinced himself that the new route inside Roscarrock was as good as the other, and that he was the man to pull it off. He surrendered his margin of safety in such small increments that he may not have appreciated the overall effect. When the engine quit, the margin of safety appeared to vanish instantly, but in reality trouble had come little by little for some time, even with the captain paying full attention. There were experienced people aboard, but for whatever reasons, they were unable to provide the support or counsel that the captain needed. Though the engine failure was a single-point equipment failure, other influences set the scene.

Sometimes monitoring is easy, as it should have been for the captain of the *Mercury*. Other times it is very demanding, perhaps to the point of being overwhelming, as it may have become for the captain of the *Maria Asumpta*. Monitoring amounts to using all the resources available—information, equipment, and people—to make navigational decisions concerning traffic, proximity to hazards, and maintaining progress. When a navigation team is in place, monitoring is a team effort. Tasks such as radio work, position fixing, radar and depth monitoring, logkeeping, and lookout can be delegated. Information can be fed to the person in charge, who then acknowledges it. (We will look at team monitoring in more detail in Chapter 12.) In a one-person wheelhouse, virtually all monitoring, with the possible exception of lookout duties, is done by the officer. Delegation is not an option.

Monitoring with a Team-of-One

In a one-person wheelhouse a watchkeeper learns to view him- or herself as a *team-of-one*, with multiple branches of responsibility that lead back to the individual in charge. Areas of responsibility include navigation, collision avoidance, communications, ship handling, looking out, logkeeping, and associated decision making. These are all important duties, so effective monitoring requires the ability to manage time, allocate energy, and prioritize. It can be helpful for the solitary watchkeeper to think of these areas of responsibility as departments, with each department requiring timely oversight and updating. Rather than staring into the night and reacting to whatever first grabs your attention, the team-of-one approach envisions a watchkeeper who moves continuously from department to department, like a supervisor, checking on the status of each area of concern. This is what the best watchkeepers do anyway, and many seasoned watchkeepers have evolved their own methodologies for accomplishing the same thing.

True, there are times when several things seem to need attention at once. But most of these situations arise because watchkeepers have either failed to manage the existing resources in the first place or they have failed to anticipate the need for additional resources. The team-of-one concept reinforces the idea that monitoring is proactive, not passive, and that all parts of it deserve attention, even on quiet days.

Boredom is a very real enemy of good monitoring. The captain of the *Mercury* may well have been a victim of boredom. While it is probably not possible to escape it entirely, the team-of-one approach can help individuals stay mentally active, making it less likely that a watchkeeper will succumb to the low levels of

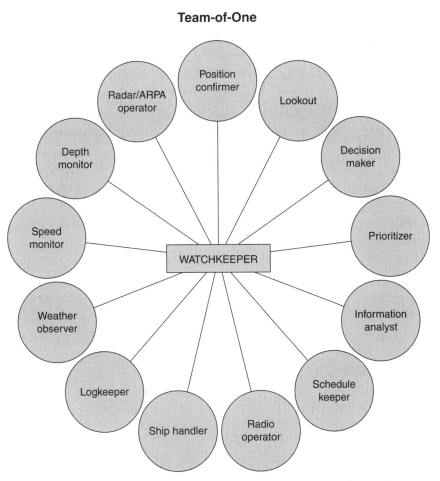

Team-of-One

The team-of-one concept means a watchkeeper has to be aware of many things at the same time. This requires an ability to shift focus between competing demands while maintaining the big picture.

situational awareness associated with boredom. An alert, engaged watchkeeper who moves between tasks has a much better chance of staying abreast of events and avoiding fixation on any one task.

Cross-Referencing

In days past, mariners spent a lot of time not really knowing where they were. Dead reckoning, that ancient method of managing uncertainty, amounted to one big educated guess. Whether offshore or inshore, navigation was labor intensive, and much of an officer's job was devoted to figuring out where the ship was. This is no longer the case. Nowadays there are few reasons not to know where a ship is. Accurate information is easily obtained via electronic instruments, at least when they are working properly, and it can be confirmed through yet other devices. The advent of these instruments has often been accompanied by complacency, laziness, boredom, and the erosion of traditional skills. Nevertheless, these "magic boxes" have their advantages and they are part of our world. A significant outcome of this evolution has been to alter the role of a navigator from someone who actively seeks information through great effort to someone who monitors and cross-references information that is readily available. In many respects, a successful watch officer must now be a shrewd analyst who continually scrutinizes data, checking to see if it is supported by other data so as to catch a trend or a malfunction. Officers who do not cross-reference assiduously are not doing their job.

Logkeeping is also part of monitoring. The logbook speaks for the watchkeeper. It is the official record of the voyage, a professional "voice" that carries far into the future. Entries should reflect a high level of professionalism by being timely, legible, and accurate. They should also be consistent and credible so that the watchkeeper doesn't wind up arguing against him- or herself at a later time. Having said this, in a choice between having a collision while maintaining an immaculate logbook or avoiding a collision with an imperfect one, the latter is generally preferable. While all aspects of monitoring are important, the ability to prioritize among them is crucial to success.

Below is a list of some monitoring and cross-referencing tasks that watchkeepers must pay attention to:

- Confirm the vessel's position in accordance with the passage plan or other procedures. This is different from glancing at an electronic chart and concluding that everything looks about right.
- Compare the primary source of position information against secondary sources.
- Compare depth-sounders to each other and to charted depths, allowing for tide.

- Compare compasses and other directional information. This is particularly important when directional information is interfaced with other electronics and has the potential to supply incorrect heading data to other critical functions such as ARPA.
- Compare radar images, looking for significant differences. Among other things, this may give clues that one of the radar units is not properly adjusted.
- Change radar range scales periodically, bumping in and out from the preferred range to see if new targets can be found. Changing scales may also require adjusting radar settings for an optimum picture. There is no point in bumping out the range scale if the existing Gain setting will not reveal anything new.
- Compare radar/ARPA displays including those pertaining to vectors, and ground and sea stabilization
- When visibility is not unlimited, use radar ranges on known objects to gauge present visibility and monitor trends.
- Compare speed over ground to speed through the water to get clues regarding set and drift.
- For the same reason, compare the course made good to the heading.
- Reassess wind and sea conditions at frequent intervals. Though weather-related information may be logged hourly, more frequent checks will keep the watchkeeper active and attuned to changes.
- Compare sources of target information: visual cues should agree with radar and AIS (automatic identification system) data as well as with intentions articulated by radio. If they don't agree, there is a malfunction or a misunderstanding.
- Check that own-ship AIS output is correct, so as not to create confusion for other vessels.
- Routinely check and adjust settings on all bridge equipment, including dimmers. Equipment may be functioning perfectly well, but if it is not adjusted correctly it will not deliver what is needed. If the VHF radio is on the wrong channel or has been turned down, the watchkeeper is not maintaining a radio watch and is deprived of information that contributes to situational awareness.
- Check standing orders, night orders, the passage plan, and any other written instructions to be sure you are operating within parameters. Reading these things once is generally not enough.
- Make timely and appropriate log entries.

*****

Good planning leads to good monitoring. The captain of the *Maria Asumpta* not only had a lot on his mind as a navigator and a sailor (and the owner!), but at the end he was improvising his passage plan. In the monitoring stage the passage plan serves as something like a collective conscience, sustaining the shared mental model around the clock. Mariners think twice about blowing off a solid passage plan. This does not mean that a watchkeeper suspends all judgment in obedience to the plan. Plans change, as indeed they must, but you can't have plan B if you don't have plan A first.

Chapter Four

Building a Passage Plan: Tactics and Tools

THE PREVIOUS TWO CHAPTERS made the case for passage planning, and we examined a conceptual framework for approaching it. But if you have never planned a passage before, and the captain has tasked you with creating a plan for an upcoming voyage, what exactly would you deliver? Clearly, it depends on the voyage. From a planning perspective, an offshore run, a coastwise transit, and an inland passage differ in significant ways. Vessel type, company policy, regulations, and the captain's preferences also matter. In this chapter we pinpoint a set of standard passage planning tools and tactics that can be adapted as needed. While it is important to have a thorough knowledge of ocean routing, our emphasis in this chapter is on coastal and confined waters because that is where the lack of effective BRM tends to exact the highest toll. More important than any particular passage planning format, checklist, or spreadsheet is the basic idea that you can avoid a world of trouble with a decent plan.

A passage plan can be organized into two main parts. The first part is establishing a *safe route* on a chart. This is largely a graphical exercise but it also involves researching many other factors that shape a safe route. The second part, sometimes known as a *passage brief*, contains everything else. The passage brief supplements charted information but is also where information relating to tides, currents, under-keel clearances (UKCs), weather, river stages, radio frequencies, and regulatory requirements is brought together. The route and the passage brief overlap and support one another in many respects, so the process of creating them is a blended effort.

One universally accepted premise of passage planning is that it should be "berth to berth." This includes open water, coastal waters, and any portion of the passage a pilot is aboard for. Mariners do not abdicate their professional responsibilities when a pilot comes aboard; to do so would invite a single-point error. Also,

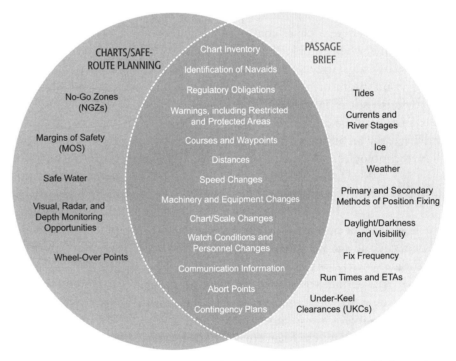

Passage planning entails acquiring and coordinating many pieces of information and putting them into a format that can be easily used by other watchstanders. Planning a safe route on the charts is one component (left), entailing a detailed review and marking of the chart. (See also sidebar Tailoring a Chart, later in this chapter.) At the far right of the diagram are the elements of a Passage Brief, which is an assembly of information critical to a safe passage, but which doesn't usually have a place on the chart. It can include weather, tide, methods of position fixing, and times of sunrise and sunset. The middle of the diagram shows other critical information that applies to both safe route planning on a chart and the Passage Brief. This information needs to be provided in one fashion or another to ensure a safe passage.

41

a passage plan must conform with company orders, standing orders, and applicable regulations. The STCW convention requires that every plan consider "all pertinent information." This is deliberately broad language that recognizes the diversity of marine operations; remember, though, that the highest purpose is to produce something that is *useful*. Thus, a passage plan should be both comprehensive and easy to use.

Route Planning: Charts

Whether paper or electronic, charts are the main medium for creating and conveying a safe route. Our goal should always be to maximize the information they provide regardless of the specific chart technology in use. The advantages of electronic charts are well established, but we should be cognizant that any technology that promotes the passive acceptance of critical information, or overreliance on a favorite source of information, is problematic for planning as well as monitoring. Whereas paper charts are comparatively cumbersome, when combined with the ability to take compass bearings, they lend themselves well to traditional visual monitoring. Since visual monitoring is so obviously essential to good watchkeeping, paper charts may still be preferable for some purposes. By and large, the techniques discussed below apply to all chart technologies.

Chart Inventory

The first order of business in preparing the route is to confirm that all the necessary charts are aboard with up-to-date corrections. Navigational charts should be of an appropriate scale for the anticipated voyage. For close-quarters work, make sure that the largest-scale chart (large scale = large detail, small area) is aboard and selected for use. Failure to use the largest-scale chart has been cited in countless incidents, including the grounding of the tug *Mercury*'s barge in Puget Sound, discussed in Chapter 3. A selection of small-scale charts (small scale = small detail, large area) should also be aboard to show the big picture and facilitate longer-range planning. In addition to the points of departure and arrival, it is prudent to carry large-scale charts for ports and anchorages along the route. Plans change and emergencies arise; sooner or later you will end up going somewhere you did not expect to go, but the obligation to navigate safely remains the same. Carrying charts and publications for other potential destinations is part of what is known as *contingency planning* (see page 53). As appropriate to the passage, other materials such as routing, climatic, load line, and pilot charts should be aboard as well. A list of navigational charts for the trip should appear in the passage brief for easy reference and so that none are overlooked during the passage.

Chart Datum

The accuracy of a charted position depends upon the accuracy of the chart on which it appears. This is partly a function of *chart datum*. (Geographic positions are based on *horizontal* chart datum, as opposed to the *vertical* chart datum used for depth and elevation.) At present, the horizontal chart datum used by the GPS system is WGS (World Geodetic Systems) 84, and most nautical charts conform. But not all. There are over 60 different datums used on nautical charts worldwide, some of which may never be converted to WGS 84 because they are based on old surveys that are impossible to reconstruct. As long as the chart you are working on is WGS 84, then GPS positions should be accurate, but the difference between WGS 84 and other datums can be truly substantial. In 2000, the *La Moure County*, a 522-foot U.S. Navy vessel on exercises in Chile tore a 40-foot gash in the bottom while carrying 240 U.S. Marines, their equipment, and vehicles to shore. The vessel was declared a total constructive loss and was later used for gunnery practice. Navigators aboard the vessel had been plotting GPS positions onto a Chilean chart that was not WGS 84. With the wrong datum, no matter how carefully they worked, they were not going to be where they thought they were. This was an expensive oversight, and by no means unique. Planners on vessels that operate exclusively within a familiar area using charts and position data that are congruent do not normally face this issue. Planners operating farther afield should use the planning stage to check and reconcile any datum differences that may exist. Most GPS receivers can readily convert most local datum but only if the receiver has been set accordingly. (Note that the chart used in the safe planning section on page 44 uses NAD1902.)

Tailoring the Chart

Official government-published charts are generic off-the-shelf tools that anyone can use for any purpose.

A very large crude carrier (VLCC) that draws 55 feet and needs half a mile to make a course change uses the same chart as a 40-foot excursion boat that can turn on a dime. Obviously, the operators of these two vessels view the watery parts of the chart very differently. Crafting the optimum route involves tailoring the chart to your needs, your vessel, and your experience. The ultimate goal is to optimize the information on the chart for the greatest chance of success. Chart annotations should be clear and should use terminology that will be clear to others. When marking charts, take care not to obscure aids to navigation or other essential details. Consult *Coast Pilots*, *Sailing Directions*, *Light Lists*, *Radio Navigation Aids*, *Local Notices to Mariners*, tide and current tables, and other standard voyage planning publications while tailoring the chart.

Aids to Navigation

An early step in tailoring a chart is to highlight all aids to navigation (navaids) along the route. Not only does this bring them to the attention of anyone else viewing the chart, but the process of identifying navaids elevates your knowledge of the waters and the marked hazards. Navaids frequently figure into formulating an optimum route, so it is a good idea to inventory these resources early in the process. Isolated hazards and radar-conspicuous objects can be treated in much the same way. Some of these may prove useful for ground stabilizing an ARPA.

You should note the distance at which lights can be seen under clear conditions, taking their geographical and nominal ranges into account. Later, if a light that should be visible isn't, it could mean several things, none of them good, and all of them important to a watch officer. While most countries note nominal ranges on their charts, some use geographical ranges or whichever is the greater of the two. As with chart datum, it is important to know what you are looking at.

In cases where the correct position of a buoy is particularly significant to a safe passage—perhaps at the edge of a shoal that you must skirt closely—establish its range and bearing from a known point in advance and note it on the chart. When making the approach, the watchkeeper can more readily confirm by radar whether the buoy is on station.

No-Go Zones

Shoal water for one vessel is navigable for another. Establishing no-go zones (NGZs) is the process of "roping off" the parts of the chart your vessel cannot pass over, under, or through. The vessel's draft and overhead and horizontal clearances are the primary restrictions, but squat and water density (salt, fresh, brackish) can also play roles in establishing NGZs. Mark these restrictions directly onto the chart with hash marks, a highlighter, or something similar. An NGZ may be an isolated hump or a large swath of submerged territory. Marking NGZs and leaving only the usable water can radically change the appearance of a chart and give a more accurate feel for the required level of navigational precision. What appears at first glance to be an expansive body of water may in fact be clogged with submerged islands or amount to little more than a creek for your vessel. Anyplace your vessel cannot physically go may as well be dry land, so why not show it as such? Establish NGZs early in the process of tailoring your chart because creating them forces a closer examination of the chart and the factors affecting your foremost concern, UKC.

For vessels that operate with a more or less constant draft, NGZs can be permanent annotations. For vessels whose draft changes from trip to trip, NGZs will have to be adjusted accordingly. Where a transit is tide-dependent, you may not be able to finalize NGZs until the execution stage, when the expected height of tide is known.

In establishing NGZs remember that actual water levels can be less than charted depths for many reasons. For example, spring low tides can result in depths below chart datum; strong winds and barometric pressure can significantly raise or lower water levels on certain bodies of water; and sea state can reduce UKCs in troughs. In large navigable lakes and inland waterways, meteorological and seasonal trends can make charted soundings unreliable for extended periods. The likelihood of debris on the bottom, siltation, and uncertainty about soundings are other considerations that go into setting up NGZs.

Opinion varies as to whether NGZs should include places that we might prefer to avoid even though the water is deep enough. Some organizations dictate minimum UKCs for their vessels, and their NGZs reflect that. Others consider anything outside a buoyed channel to be an NGZ. The problem with this is that should your vessel, or an oncoming ship, experience a steering failure, an engine failure, or some other emergency, you may be compelled to consider entering water that was not intended. If a place that is physically navigable has been designated an NGZ for reasons other than physical constraints, it could

44   Chapter Four

Tailoring a Chart

The accompanying chart (a segment of Chart 14909, 1:80,000, North American Datum 1902) has been annotated for the transit of a 135-foot schooner drawing 13 feet, bound through the Porte des Morts Passage between Washington Island and the Door Peninsula in Wisconsin on Lake Michigan. (Chart courtesy NOAA)

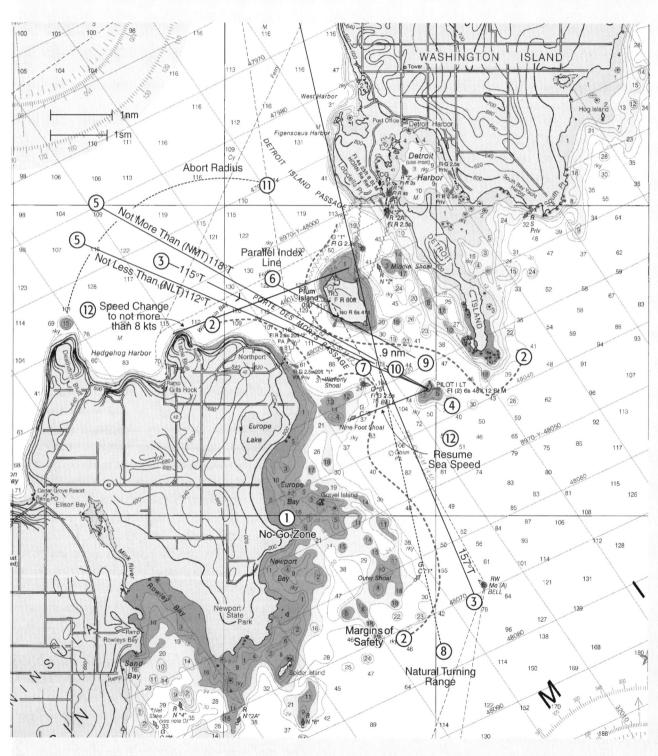

① **No-Go Zones (darkened areas).** The captain has established a minimum UKC of 5 feet for these waters under normal operating conditions. Accordingly, no-go zones (NGZs) have been drawn in (highlighter works well for this too) along the 18-foot depth contour. These NGZs significantly expand the hazards represented by Plum and Pilot Islands. The NGZ joins Waverly Shoal and Nine Foot Shoal into a single, large sunken island. Note that water between Waverly Shoal and the green '5' bell buoy is shown as navigable if necessary though it falls outside the buoy. In a heavy chop, larger UKCs may be necessary.

② **Margins of Safety (dotted lines).** In the confined water of Porte des Morts Passage, a margin of safety (MOS) of 0.3 nm has been chosen to provide a buffer between safe water and the no-go zone. This buffer can be used to provide reaction time if a vessel strays from the track line, or provide sea room to deal with unexpected difficulties.

South of Nine Foot Shoal, the MOS is expanded to half a mile on account of the irregular, unmarked shoals extending from shore, and the fact that there is more sea room to spare.

③ **Track Lines.** The first track line (115°T) has been established to take advantage of various monitoring opportunities. The second track line (157°T) is set up to pass 0.2 nm west of the Red and White Morse Alpha bell buoy stationed east of Newport Bay. This track line passes close enough to the buoy to confirm your position under most circumstances, but far enough away to avoid at least some of the traffic that may be using the buoy for a waypoint. Using buoys as waypoints has resulted in numerous allisions when distracted or inattentive watchkeepers on autopilot neglected to avoid the buoy at the end of the leg.

④ **Head (or Stern) Marks.** The course for transiting Porte des Morts Passage is chosen to take advantage of Pilot Island as both a visual and radar conspicuous head mark.

⑤ **Clearing Bearings.** Clearing bearings on Pilot Island can be used to keep the vessel in safe water and out of the MOS. If a bearing on Pilot Island exceeds (NMT) 118°T, then the vessel is at risk of entering its MOS to the north. If a bearing on Pilot Island is not less than (NLT) 112°T, then the vessel is at risk of crossing into its MOS to the south.

⑥ **Parallel Indexing.** A parallel index line tangent to Plum Island can be used to monitor cross-track error on the approach to Pilot Island and also to monitor proximity to the 19-foot spot southeast of Plum Island.

⑦ **Beam Bearing.** A beam bearing on the green '5' bell buoy near Waverly Shoal can be used to provide warning of the approaching course change. An offset VRM and EBL on Pilot Island can be used to confirm that the buoy is on station.

⑧ **Natural Turning Range (LOP).** The turn to the south (157°T) could be accomplished safely at any number of points; however, a natural range formed by the east side of Plum Island and the southwest lobe of Washington Island makes for a conspicuous turning point. On the next leg this natural range can serve as a clearing bearing to warn a watchkeeper that the vessel is approaching Outer Shoal.

⑨ **VRM Turning Range (Distance).** A distance off Pilot Island can be used as a turning range for the course change to 157°T. With Pilot Island dead ahead, a VRM set at 0.9 nm will become tangent to the island about the same time the vessel reaches the waypoint, and the natural range described above (Plum Island and Washington Island) aligns.

⑩ **Depth Verification Point.** The waypoint northwest of Pilot Island was established in conjunction with a natural range, a head mark, and a turning range. However, it happens to coincide with a sounding of 103 feet. This is useful information for both planning and monitoring.

⑪ **Abort (Radius) Point.** An abort radius has been drawn 3 miles seaward of where the MOS begins to converge, and safe water reaches its narrowest point. Options for dealing with unexpected problems are much reduced once the vessel has entered the passage. Three miles may seem premature for some operators but the approach is not well suited to anchoring, thus further restricting options in the event of a mechanical failure.

⑫ **Speed Change.** Reduce speed to not more than 8 knots upon crossing a line extending from the north tip of Plum Island and Table Bluff to the west. Resume sea speed upon crossing a line between Pilot Island and the green can '3' on Nine Foot Shoal.

create confusion in a high-pressure situation and restrict options unnecessarily. Strictly speaking, NGZs are based on physical constraints and are not merely areas you prefer to avoid.

Restricted, Protected, or Environmentally Sensitive Areas

Safe routing involves identifying restricted areas that impinge on the route. These may pertain to maritime security, protection of the environment, military activity, fisheries, and so forth. Think of these areas as regulatory NGZs, even if they are otherwise navigable. You are legally obligated to honor these restrictions, but do not confuse them with the physical constraints associated with NGZs.

After the safety of life, protection of the environment has evolved into the mariner's second biggest duty under the law. The standard of care for this is very high. The Manila Amendments to the STCW convention contain the following language under the section entitled Watchkeeping at Sea: Protection of the Marine Environment:

> The master, officers and ratings shall be aware of the serious effects of operational or accidental pollution of the marine environment and shall take all possible precautions to prevent such pollution, particularly within the framework of relevant international and port regulations.

When it is necessary to pass near areas that are protected or are known to be environmentally sensitive, the plan should take this into account, and the chart should highlight such areas.

Margins of Safety

Margins of safety (MOS) apply to all sorts of operational choices, including the route. Every case we have examined thus far involved insufficient margins of safety of one type or another. Mariners should never operate at the very edge of feasibility, and good planning in all its forms helps keep this from happening. In the context of route planning, a MOS is a navigational boundary that designates where you don't want to go, but where you could go if you had to. A MOS is usually related to UKC, but there may be other factors. The MOS is established *before* you make track lines, because the analytical process behind a MOS helps determine, eventually, the best placement of the track lines. Crossing over a MOS is not necessarily cause for panic, but it does mean that you are closer to danger than planned. Thus warned, you can make decisions and take corrective action based on defined, rather than arbitrary, parameters.

Unlike NGZs, which are nonnegotiable physical restrictions, MOS are subjective. Within the natural constraints of the situation, you can *choose* how much buffer to have. A useful way to formulate a MOS is to review the consequences of something going wrong at a particular point. For instance, lee shores are always a potential hazard, but a lee shore that is also a rocky cliff facing open ocean is different from a lee shore that is a mudflat in protected waters and where navigational choices are clearly limited. A lee shore for a tugboat towing 4 million gallons of heating oil is different than it is for a recreational boater out for a Sunday sail. If the potential consequences are severe, then you should have a large MOS. One test of a sound or useful MOS is to consider how you would explain how you arrived at it to another officer in a planning conference. Typical considerations in establishing an MOS include the following:

Maneuverability: A highly maneuverable vessel may be able to justify a smaller MOS than a vessel that requires more turning room. The *Maria Asumpta* was not very maneuverable and should have had a much larger MOS.

Size of vessel or length of tow: It is possible for the GPS receiver to show your vessel in good water while the far end of a long tow is in a different situation. In Puget Sound it was the *Mercury*'s tow that took the bottom, not the *Mercury* herself.

Speed: A vessel traveling at 20 knots needs more advance notice of hazards than one traveling at 5 knots. On the other hand, a fast vessel can get out of trouble more quickly, and slower vessels, such as the *Mercury*, are more influenced by current.

Weather: Some things haven't changed since the age of sail. As mentioned above, operating on a lee shore requires a greater MOS than operating with an offshore breeze. Engine failures and other emergencies are a question of when, not if. Both the *Maria Asumpta* and the tug *Scandia* were on lee shores with small margins of safety. Additionally, poor weather poses greater hazards than fair weather. The *Scandia* used a fair-weather MOS for a foul-weather voyage.

Visibility: All things being equal, it is harder to monitor events in restricted visibility and in darkness. This makes a larger MOS prudent under those conditions.

Traffic density and patterns: Interactions with other vessels are one of the chief reasons for departing from the track line. In selecting an MOS you should consider how traffic will impact your ability to maintain the track line and how much deviation you can tolerate.

Experience: MOS can, and should, be adjusted to the level of crew experience. Putting aside the risk of complacency for the moment, less experience generally argues for a larger MOS.

Accuracy and reliability of navigational tools and methods: There is no piece of navigational equipment that is ideal for all needs. There are times when radar is preferable to GPS, and when visual bearings are preferable to either of those. In many parts of the world, charts are based on scanty, dated surveys or are not available in the best scale (remember the *Safari Spirit?*). These things affect a watchkeeper's ability to monitor with certainty, and the MOS should reflect that. When more time-consuming methods (e.g., visual bearings) are necessary, the MOS should take that into account, too.

Monitoring opportunities: If we are to view navigation as something more than GPS-based, then our ability to know where we are is also related to the availability of monitoring opportunities. The paucity or abundance of navaids and natural monitoring opportunities (prominent radar targets or visual features) influences the MOS.

Expecting the unexpected: The underlying purpose of a MOS is to create a buffer that buys you time and sea room to manage unexpected events. The MOS are a visual demarcation of what we already know to be true about working on the water: expect the unexpected.

As you can see, appropriate MOS vary with circumstances. Coming down the coast in daylight with good visibility and an offshore breeze, it may be acceptable to pass half a mile from shore and catch a favorable current. At night, in restricted visibility, with a 30-knot onshore wind and a new watch officer at the conn, you need a larger MOS, even if it costs some time and fuel. Adjusting the MOS for the situation is an example of the active engagement that good passage planning is intended to encourage.

It is always preferable to establish MOS in a manner that makes them easy to monitor, either by sight, radar, or other means. Coasts with few prominent features offer fewer options for monitoring an MOS. In narrow waterways, where no-go zones start at the riverbank, a MOS may not even be a viable concept. In general, MOS work best where you have navigational *choices*.

Safe Water

What remains on the chart after the NGZs and MOS have been added is *safe water*. This is where waypoints and track lines go. Though the plan is to follow the track line, deviations are inevitable. All safe water is fair game for normal operations, such as course changes for traffic.

Crafting the Optimum Route

Having highlighted points of interest and redrawn the chart to suit your vessel and experience, now reexamine it to determine the optimum route. Efficiency is measured in more than one way, and the optimum route is not necessarily the shortest distance between two points. While saving time and fuel is a central consideration, arriving intact is also a form of efficiency, so look at other factors too. Electronic charts are immensely useful, yet overreliance on easy information has led to many accidents and will lead to many more. Rather than relying solely on a computer-generated optimum route, scrutinize the chart closely for any visual, radar, and depth cues along the way that you can use to monitor progress and cross-reference other sources of information. An optimum route does not always transit the deepest water or pass equidistant between two hazards. For instance, when a natural range is available, it may make sense to trade some distance or UKC for the opportunity to monitor the track line with the confidence and precision that a natural range affords. While you could take advantage of these opportunities on the spur of the moment, building them into the plan makes it more likely that they will be applied by *all* watchkeepers instead of just some.

Establishing the optimum route takes time and thought, but such things are expected of professionals. Your planning needs to prevent only one career-ending incident to be worth the trouble.

Position-Monitoring Tools

In the world of small vessels, you don't always have the pleasant option of running a route that lies between red and green flashing buoys while steering on a lit range in a dredged channel with a pilot aboard. Even when you do, any tool that brings peace of mind and certainty to a watchkeeper is a good tool. Many of the position-monitoring techniques that you ultimately employ underway you should first apply when planning the route. The following is a brief catalog of well-worn yet frequently overlooked position-monitoring tools that can be used in conjunction with the chart in the planning stage. This is old-school navigation, and such tricks of the trade have kept mariners out of trouble for years. Annotating authorized charts to include additional local features is an ancient practice that can help with passage monitoring. Such additions may not constitute "official" information, but when used correctly they can constitute smart piloting. When visual references are close by, take care to monitor them from the same spot in the wheelhouse or on deck to avoid inconsistent readings.

Head (or Stern) Marks

A head or stern mark is a special-purpose bearing on an object that is in-line with the desired track. Absent current or leeway, the head mark will be dead ahead (or the stern mark directly astern) when the vessel is on its track and steering its prescribed course. If it is not, then it may indicate the presence of *cross-track error*—the amount by which a vessel is left or right of track. Gauging cross-track error is critical to decisions about maintaining the track line, so any means of assessing cross-track error is hugely important to voyage monitoring.

When steering a vessel to counteract current or leeway, a head mark may not be dead ahead even though the vessel is on its track; in such a case the head mark should maintain a constant bearing even though it is not dead ahead.

Beam Bearings

A beam bearing is a special-purpose bearing used to indicate a vessel's progress *along* its track. If the vessel is on its prescribed heading, a charted object on the beam gives an instant visual cue as to progress fore and aft. Beam bearings are often used to mark the approach of a waypoint or wheel-over point (see below). Like any bearing, a beam bearing is not a fix, only a line of position. Unless paired with a distance off, as described below, or some other information source, a beam bearing does not tell you whether the vessel is left or right of its track, which is critical to know before making a turn. The beam-bearing relationship changes if a vessel is not steering its prescribed course, and this must be taken into account if a vessel is unable to steer the planned course.

The rate at which a beam bearing changes can be useful for gauging speed over the bottom when you need to get a feel for whether the vessel is moving too fast or too slow for the situation, such as when approaching a dock or lock.

Clearing (or Danger) Bearings

A clearing (or danger) bearing is a special-purpose bearing used to monitor a vessel's proximity to an area of interest. Clearing bearings are useful in monitoring the approach of a no-go zone or a margin of safety. It is good practice to label clearing bearings "not less than" (NLT) or "not more than" (NMT), followed by a three-digit bearing. This enables anyone on the bridge to monitor the trend toward or away from an area of interest simply by observing the change in the bearing. A clearing bearing is only a line of position, but it can indicate whether the vessel is standing into danger.

Natural Ranges

A natural range is formed by the alignment of any two fixed objects, one nearer and one farther, whose positions are accurately charted. Since a natural range is a unique relationship between two references, it is among the most foolproof methods available for establishing a line of position—you're either on it or you're not—which makes it highly desirable. The utility of ranges is borne out by the fact that man-made ones appear worldwide. Where they occur naturally they should be incorporated into route planning when practical. You can use a natural range to establish a line of position for any purpose, including all the uses of the special-purpose bearings discussed above.

Ranges have an additional watchkeeping function in that they provide excellent opportunities to check for compass error. For this reason, too, you should note natural range opportunities on the chart.

Parallel Indexing and Clearing Ranges

Parallel indexing is among the most versatile radar tools for monitoring cross-track error. You can also use parallel index lines to designate a clearing range (danger range), which is the minimum distance off a radar-conspicuous object that must be maintained to stay in safe water. The ability afforded by parallel indexing to monitor trends at a glance not only helps watchkeepers stay abreast of the vessel's position, but it also can facilitate decisions about whether it is imperative to regain the track line immediately or whether that action can be postponed in order to accomplish some other priority, such as avoiding traffic. While you can monitor a position by other means, the beauty of parallel indexing is that it puts measurable, real-time information on cross-track error into a visual context that is radar based, and therefore independent of other sources.

Parallel indexing has obvious benefits in restricted visibility but it is also a great everyday tool. Like other radar techniques, you should practice it in clear weather so that you are comfortable with it in poor visibility. Unfortunately, many smaller vessels do not have radars equipped with a dedicated parallel index feature, but an offset electronic bearing line (EBL) and a variable range marker (VRM) are standard features that can be configured to produce a "poor man's" parallel index feature. Note that parallel indexing should be used with great caution in heads-up mode as it may not provide a realistic view of trends affecting the vessel.

Turning Ranges

A turning range is a distance off a known object that is used to indicate where a course change should occur. You can also use it to monitor the course change once it has begun. By locating your track line to take advantage of a radar-conspicuous object dead ahead (or dead astern), you can use a simple VRM centered on the vessel and set to the desired distance off to indicate the approach of the turn. As the VRM draws tangent to the radar-conspicuous object ahead (or astern), it indicates that you have arrived at the turning point. Rate of turn can be modified to maintain the distance off as desired. A parallel index line set perpendicular to your course can also be used to establish a turning range. As the vessel approaches the turn, the line makes contact with the edge of the chosen radar target.

Variable Range Markers and Electronic Bearing Lines

This is not the place to enumerate the many creative uses for VRMs and EBLs for voyage monitoring and collision avoidance, but in the interest of maximizing your resources, remember that even fairly basic, unstabilized radars provide these features. In particular, you can use the offset features of a VRM and an EBL to greatly expand the capability of a simple radar. As with other techniques discussed above, your route planning should consider how you can apply these basic tools for the best result.

Depth Verification Points

Depth has been described by some as a vertical line of position. In most places, depth changes too gradually to ascribe it the level of the precision that we normally associate with LOPs. Still, the concept of using depth to confirm your position is thoroughly valid. In many places a chart will show an isolated spot—a sill, a trench, or some other well-defined bathymetric feature—where the depth changes markedly. Since good monitoring entails constantly cross-checking information, depth has value beyond simply noting when you are or aren't about to go aground. This is not to suggest that we should build an entire route around transiting waters with distinctive soundings. But if a well-planned route passes through an area with well-defined bottom features, you should note these on the chart and use them to confirm not only your approximate position but also the accuracy of the depth-sounder.

The handful of navigational techniques reviewed above could be greatly expanded upon with variations. The main point is to remember that any trick of the trade that has value to a navigator also has value to a passage plan. Rather than viewing navigation as something that starts upon casting off, we should view it as a skill that is *first* applied in the planning stage. You do not and cannot apply all techniques at once, but each has its place sooner or later.

Track Lines and Waypoints

Track lines and waypoints are interdependent—the location of one determines the location of the other. Laying a track is perhaps the most elementary aspect of voyage planning, yet strangely enough they are frequently omitted by mariners of all stripes, sometimes with disastrous results. The captain of the *Mercury* was not using one when his barge grounded. While it hardly seems necessary to explain its importance, it must be said that the track is where you want to go. It is the baseline for the voyage, and some thought should go into that. Failure to lay a track line deprives you of the continuous ability to measure your position against a meaningful standard.

The overriding consideration when establishing a track line is choosing a safe route that you can monitor with confidence. In open water, where navigational constraints are minimal, you typically select track lines and waypoints to yield the shortest distance between two points. In confined waters you should become adept at setting up tracks that take advantage of the natural monitoring opportunities found along the route. Since cross-track error is perhaps the most critical metric of performance with regard to maintaining a track, arrange your tracks to make use of head marks, stern marks, natural ranges, and parallel indexing whenever practical.

For each leg, you should identify the minimum anticipated UKC, highlight it, and enter it into the passage brief. Later you can use it to monitor areas where UKC is of greatest concern. Courses should be clearly indicated, in true, magnetic, or compass in accordance with the prevailing source aboard. Keep in mind that transposing course digits is among the most common errors, both verbally and in writing.

Sometimes you will operate in places where rigid track lines are virtually impossible to follow. This may be due to traffic patterns or waterways that change orientation so frequently that the vessel never steadies up on one course for long. Instead you rely upon visual (or radar) cues to make appropriate adjustments until you can settle on a steady course again. In such situations, does laying a track serve any practical purpose? Remember, no one ever plans to become disoriented, but it does happen. Even in good visibility, a misidentified buoy can lead a person to alter course toward shoal water inadvertently. In fog or darkness opportunity for disorientation is even greater. Therefore, even when following a track line appears impractical or unnecessary, having one there can provide a reference point that may come in handy when least expected.

Also, compasses have been known to drift gradually or fail outright. When this happens, as it will, it is the watch officer's job to catch it. On a bridge where compass input is fed to other electronics, wrong information can be transmitted to multiple points, in effect spreading the lie. A charted track line can help alert you to a discrepancy between what the compass reads and where the vessel is actually heading.

The need to confirm your position is often greatest immediately before and after a course change. Natural ranges, beam bearings, and turning ranges can all provide useful information when approaching a waypoint. You can start thinking about this as you approach a turn, or you can locate waypoints to take such things in account during the planning process. It is not always possible to anticipate the environmental factors at the time of route planning, so wind, current, traffic, and other factors may require you to adjust a planned waypoint.

Waypoints are most often used in connection with course changes, but they can also be used to indicate other changes pertaining to machinery status, speed change, calling additional hands, watch conditions, readying anchors, pilot embarkation, VHF channels to monitor, or any other shift of operational posture. If these things are marked on the chart, they are less likely to be overlooked.

Wheel-Over Points

Wheel-over points are akin to waypoints in that they represent a geographic place where something must happen. Specifically, wheel-over points indicate where a vessel must begin its turn in order to complete the course change at the waypoint. Wheel-over points are based on advance, transfer, squat, and other maneuvering characteristics. On large ships, maneuvering characteristics appear on a bridge poster, but many smaller vessels do not have bridge posters which means that wheel-over points must be based on experience. Marking wheel-over points on a chart is standard procedure aboard merchant ships but is not always embraced aboard smaller, more maneuverable vessels. This is perhaps understandable as smaller vessels may have more clearance than larger ones and since they are often more responsive to rate-of-turn adjustments, less planning may be needed. However, many small vessels aren't that small, especially when joined to a tow. Furthermore, smaller vessels frequently operate in smaller bodies of water, resulting in the same small tolerances that make wheel-over points useful to larger vessels. In any event, the annals of groundings, including the *Exxon Valdez*, contain no shortage of ships that turned too soon or too late. Marking a wheel-over point on a chart draws attention to the importance of the maneuver and can help ensure that the vessel goes where it is supposed to.

Like waypoints, wheel-over points should take advantage of natural monitoring opportunities such as beam bearings, turning ranges, and so forth. The same factors—wind, current, and traffic—that can force a change of waypoint can force a change of wheel-over point.

Beginning a turn in the right place is one thing; finishing it in the right place is another. Where possible, accompany turning marks with a reference point for the next leg to steady up on. This could take the form of a head mark, a stern mark, a natural range, or a parallel index line. If the captain of the *Safari Spirit* had done this, his error would have been immediately obvious.

Distances

Distances for each leg should appear either on the chart, in the passage brief, or both. Doing it once, carefully and correctly, can save time later and minimize the chance of mistakes arising from a hasty and incorrect measurement under pressure. The captain of the *Safari Spirit* made a faulty distance measurement in a different context from what we are discussing here, but his error illustrates the importance of getting it right, and how easy it is to get it wrong.

Additional Notes for the Chart

Once you have crafted the optimum route, it is time to jot down any additional warnings, notes, or prompts that will help the watch officer stay on top of what needs to happen. Chart notes may consist of nothing more than the captain scrawling, "Call me here," or "Switch to Channel 12 here," or they may involve more detail that sends the watchkeeper to some other source of information (e.g., publications) to get the whole story. Chart notes maximize the benefits of planning by keeping people dialed into the plan, and they facilitate teamwork by putting information where all can see it. The following are some examples of useful chart notes that may also be addressed by the passage brief.

Speed Changes

Speed is as much a part of passage planning as it is of monitoring. The factors that go into establishing a safe and reasonable operating speed will be examined on page 56. Speed changes may appear in the passage brief, but they can also go onto the chart. This may be especially helpful when there is a gradual transition from open water to confined water. Without a speed change prompt, it is conceivable that a distracted watchkeeper might carry cruising speed too long. Other speed changes are mandated by local regulations; failing to comply creates a violation.

Chart and Scale Changes

Any transition is fertile territory for human error, and moving between paper charts is no exception. It is all too easy to make a mistake when transferring the last known position from the old chart to the new one, particularly if the new chart has a different scale and the new scale is misinterpreted. When navigational hazards are just beyond the edge of the present chart, the unwary watch officer has little warning as to what lurks on the near edge of the next chart. To help with forward thinking, it is customary to indicate on each paper chart where the next chart picks up. When navigating with paper charts, you should keep the number of charts that are out at one time to a minimum. When multiple charts are piled on top of one another, it is easy to take measurements or information from one and apply it to another incorrectly. Differences in the way latitude and longitude increments are displayed can be overlooked. These kinds of operator errors can also happen when charts contain insets of a larger scale than the main chart. With paper charts, conventional wisdom dictates that you always use the largest scale, though there can be exceptions to this generally sound advice.

A clear advantage of electronic charts is that you don't have to deal with abrupt transitions between charts but they are not free from opportunities for misinterpretation. Though electronic charts have a capacity for magnifying detail and changing scale, much like forgetting to reduce speed in a timely fashion, watchkeepers are susceptible to sticking with an inappropriate scale, and in the case of vector charts, an inappropriate level of detail, longer than they should. Situational awareness is the primary defense against this sort of lapse, but a notation reminding the watchkeeper to check scales at critical junctures can help avoid such an oversight.

Machinery and Equipment Readiness

In the normal course of a voyage, the readiness of certain machinery, systems, and equipment changes with the vessel's location. For instance, you don't need a bow thruster warmed up and ready to go 20 miles offshore, but it had better be ready in time for docking. Somewhere in between, someone needs to make sure that this happens. Establishing just when and where that change of status should occur is part of passage planning, and a chart note can help alert the watchkeeper to these needs. Timely notice is especially important if the action requires coordination with other personnel. Many smaller vessels do not have manned engine rooms or a duty engineer awake at all times. Frequently, the engine department is a team-of-one, perhaps assisted by a mechanically inclined deckhand. Some small vessels have no dedicated engineer at all. Instead, the captain or the mate is the engineer. Under these circumstances, deck officers must be especially proactive and forward thinking in coordinating machinery needs with the passage plan. Below are some examples of machinery or equipment tasks that often require coordination between deck and engine crew:

- Testing the gear and thrusters upon approaching docks, locks, or confined waters.
- Lighting off a backup generator, steering pump, etc., upon approaching a critical stage in the voyage.

- Switching to high suction for cooling water if there is a likelihood of clogging the cooling system with silt or debris, causing the system to overload and black out the ship.
- Switching to low suction for cooling water if there is a likelihood of clogging the cooling system with ice, flotsam, or other surface debris.
- Switching to full fuel tanks upon approaching confined waters or heavy weather, or when crossing a bar or a similar situation where there is a likelihood of kicking up tank sludge or otherwise jeopardizing the fuel supply at an unforgiving time.
- For the deck crew, readying windlasses, winches, docklines, fendering, lights, and embarkation ladders as needed.

As much as possible, the safety of a boat should not be completely built around one individual remembering a critical duty in time, as memory alone is an unreliable defense against human error. If the passage plan anticipates a change of machinery status that impacts safe navigation, then it should appear on the chart or in the passage brief, or both.

Watch Conditions and Crew Readiness

An overriding consideration in determining the composition of a bridge team is the ability to maintain a proper lookout at all times, no matter what is happening. This means that sometimes you need a dedicated lookout and other times one may not be enough. Sometimes you cannot maintain adequate situational awareness without additional assistance with monitoring radar, the radio, or the vessel's position. In other words, you need a bridge team. On other occasions you need extra hands for docking, anchoring, or changing the towing configuration or the sail plan. As noted earlier, sometimes you want the engineer standing by in the engine room to monitor machinery and respond to any needs. When you can anticipate manning adjustments such as these in the planning stage, and you very often can, a note on the chart will prompt whoever is on watch to initiate these things in a timely fashion rather than at the last minute, or too late.

Regulatory Obligations and Warnings

Some regulatory obligations, such as those pertaining to lifesaving equipment, are fairly permanent for a given vessel on a given route. Other regulatory obligations depend on the vessel's location. Failure to comply with these regulations for any reason can result in significant fines and other serious consequences. Here are some examples of localized regulations that can impact a passage, and therefore the passage plan:

- Discharge restrictions
- Speed restrictions
- Mandatory reporting systems pertaining to VTS, maritime security, or environmental protection
- Restricted zones relating to maritime security, law enforcement, military activity, etc.
- Mandatory monitoring of specified VHF channels
- Pilotage requirements
- Equipment tests
- Display of mandatory flags or signals

Just as ignorance is no excuse for not obeying the law, neither is forgetfulness. While the details of a particular obligation may warrant close study, a chart notation can help avoid an unintentional violation. If the situation involves consulting a publication, it can be helpful to bookmark and highlight essential parts of it so that watchkeepers can go straight to the source. Special warnings that come in via navtex or some other source in the course of the transit should incorporated into the passage plan and passed along.

VHF Monitoring

Part of watchkeeping is maintaining a radio watch and a radio presence. If your radios are not watching as they should, then you have a blind spot. Chart notes should indicate where certain VHF channels should be monitored and where certain calls should be made, including Sécurité calls.

Abort Points

Boats are built to move. Everything about working on boats is related to getting where you're going and doing what you've got to do. In consequence, one of the hardest decisions to make is the decision to abort a voyage. It might look bad. It might cost the company money. Everyone knows that working on the water entails accommodating powerful forces beyond anyone's control, and that sometimes turning back is absolutely the right call. Nevertheless, an even more powerful collective sense of duty and momentum sometimes drives people to press on when they shouldn't.

An abort point is a self-imposed trip wire designed to trigger a reassessment of the navigational situation. It represents a point of no return, a last, best chance to turn around before committing the vessel to a high-risk situation. Any condition that jeopardizes a safe passage can be a reason to abort a passage. If the conditions on which the passage plan is premised are not present, then proceeding is a violation of the plan. An abort point would be appropriate for a transit that is premised on specific tide, current, weather, visibility, river stage, or bar conditions. If the prevailing conditions don't mesh with the specified plan, then there is a clear rationale for a decision to abort. Abort points might be established in conjunction with a machinery/equipment status upon entering waters where there are no anchorages, tie-up spots, or turning basins, or no means of reversing course. One clue that an abort point is in order is that margins of safety begin to converge, leaving minimal safe water to work with. Abort points could be related to security and crew safety. While the unexpected can still occur, an abort point prompts you to make a conscious effort to ensure all is well before proceeding. For the captain of the *Scandia*, an abort point could have prompted a reassessment of his intention to leave protected waters with a jury-rigged anchoring arrangement and a storm bearing down on a lee shore. For the captain of the *Maria Asumpta*, an abort point, in conjunction with margins of safety, could have helped him realize that he was standing into danger much earlier than he did. For the captain of the *Anne Holly*, an abort point premised on having daylight, assist boats, or both, might have helped him make a better decision about transiting the Eads Bridge. Indeed, when asked what he thought could best prevent a repetition of the accident, the captain replied, "I'd like for them to put a restriction on these bridges northbound as well as southbound. . . . It would be a lot safer if everybody could run in the daylight." Such a recommendation amounts to an abort point: if it's night, you don't go. In all three cases, a sense of momentum was at work, but an abort point had the potential to act as a bulwark against it. In all three cases a more advanced approach to appraisal and planning was needed in order for abort points to even be considered.

Planning is the opposite of spontaneity. Abort points bring focus and transparency to high-risk decisions that have a modest upside (keeping the schedule) versus a massive downside (an oil spill, injury, death, etc.). In one sense, the concept of abort points can be seen as impinging upon an officer's or captain's discretion. On the other hand, abort points provide a clear rationale for not proceeding when there might otherwise be pressure to carry on. Like so many aspects of voyage planning, abort points serve as a reality check against wishful thinking and marginal decisions.

Contingency Planning

Contingency planning is the act of developing viable alternatives in advance in case Plan A is aborted. Since dealing with the unexpected is a normal part of vessel operations, it stands to reason that if you have a plan, then you need a contingency plan, too. Knowing your options is just common sense on boats.

There are at least two kinds of contingency plans. The first involves *delaying* until circumstances are better: waiting for traffic to clear, fog to lift, daylight to come, the weather to change, or a critical repair to be completed. In such cases a contingency plan may be nothing more than staying put, heaving-to, anchoring, or tying up temporarily. This kind of contingency planning involves ascertaining the availability of maneuvering room, anchorages, and emergency berths. A valid contingency plan for both the *Anne Holly* and the *Scandia* would have been staying put until one or more factors turned in their favor.

A second type of contingency plan involves creating an *alternate route* to your destination. The need for this can arise due to an unexpected security restriction, weather, an emergency, or an obstruction of some sort. Both kinds of contingency planning should be noted on the chart if practical.

Contingency planning is an exercise in expecting the unexpected—within reason. Boats and crews are mobile, which means that, theoretically, they can be diverted to new destinations at any time. This can be a business decision or it can be in response to an emergency, but every vessel should be equipped and operated with an eye to these possibilities. When plans change, the passage plan must be modified, and sometimes a new one must be created from scratch. However, there is an element of probability that shapes this kind of contingency planning as we can never be prepared for all possibilities. Cobbling together a contingency plan on the fly is possible, and sometimes unavoidable, but that kind of improvisation is also prone to error. One of the first steps in the formation of an error chain is often a change of plan; therefore a measure of contingency planning should be part of passage planning.

The Passage Brief

Having examined some tactics and tools for developing a safe route, we turn to the passage brief, where other

PASSAGE PLAN

VESSEL:	CARGO:		24-HOUR Weather Forecast:	
FROM:	TO:		DATE:	MASTER:
ETD/DEPARTURE:	ETA/ARRIVAL:			PAGE: 1 of 1
TIDES:		Project Channel Depth	POSITION RECORDING METHOD: PAPER CHART	DRAFT: FWD: AFT: MEAN:	CHARTS:

Wypt #	FROM	TO	GYRO HDG	DIST ON LEG	AVG SPEED	DIST TO GO	ETA TIME DATE	CONTROLLING DEPTHS MIN. U.K.C.	PREFERRED FIX METHOD	NOTES — CHARTS USING, SPECIAL INSTRUCTIONS, HAZARDS, SPECIAL AREAS
1		Time Completed:				0.0			VISUAL, RADAR	
2		Time Completed:				0.0			VISUAL, RADAR	
3		Time Completed:				0.0			VISUAL, RADAR	
4		Time Completed:				0.0			VISUAL, RADAR	
5		Time Completed:				0.0			VISUAL, RADAR	
6		Time Completed:				0.0			VISUAL, RADAR	
7		Time Completed:				0.0			VISUAL, RADAR	
8		Time Completed:				0.0			VISUAL, RADAR	
9		Time Completed:				0.0			VISUAL, RADAR	

PILOTS:
CONTACT ON CH:
OTHER CONTACT:
TIME OF CONTACT:

TOTAL DISTANCE: 0.0
DISTANCE FOR PAGE: 0.0
AVERAGE SPEED:
E.T.A.:

PREFERRED FIX METHODS: VISUAL, RADAR, GPS
TIDAL CURRENT: FLOOD, SLACK, EBB

PREPARED BY: CHECKED BY:

Water Density: SW____ FW____ Brackish____
Assistance Req'd Yes____ No____
Pilot Req'd Yes____ No____
Departure Restrictions: ____
Arrival Restrictions: ____

Narrows to Bolivar Roads

NOT FOR NAVIGATION

Leg to	Bearing	Speed	Turn	Distance	ETA (TTG)	End Position
104 D IMTT	068.1° T 081.1° M	6.00 kn	24° to starboard	0.670 nm	07:06 (7 mins)	40°38.846'N 074°06.641'W
104 C	092.4° T 105.4° M	7.00 kn	28° to port	0.655 nm	07:12 (6 mins)	40°38.818'N 074°05.777'W
100 B	064.1° T 077.1° M	7.00 kn	9° to starboard	0.503 nm	07:16 (4 mins)	40°39.038'N 074°05.182'W
104 A Con Hook	074.1° T 087.1° M	7.00 kn	31° to starboard	0.221 nm	07:18 (2 mins)	40°39.098'N 074°04.901'W
104 RW "KV"	106.0° T 119.0° M	7.00 kn	51° to starboard	0.787 nm	07:25 (7 mins)	40°38.881'N 074°03.904'W
103 Narrows	157.6° T 170.7° M	10.8 kn	10° to starboard	2.630 nm	07:39 (15 mins)	40°36.449'N 074°02.586'W
A Ambr 21-22	167.7° T 180.7° M	10.8 kn	0°	2.512 nm	07:53 (14 mins)	40°33.995'N 074°01.881'W
Ambr 17-18	168.0° T 181.0° M	10.8 kn	23° to port	1.737 nm	08:03 (10 mins)	40°32.296'N 074°01.406'W
Ambr 13-14	144.6° T 157.7° M	10.8 kn	27° to port	0.904 nm	08:08 (5 mins)	40°31.559'N 074°00.718'W
100 Ambr 3 & 4	117.1° T 130.1° M	10.8 kn	42° to starboard	5.536 nm	08:39 (31 mins)	40°29.036'N 073°54.237'W
Ambr G "1A"	159.2° T 172.3° M	10.8 kn	25° to starboard	0.963 nm	08:44 (5 mins)	40°28.136'N 073°53.788'W
Shrewsbury	184.3° T 197.4° M	10.8 kn	0°	7.708 nm	09:27 (43 mins)	40°20.450'N 073°54.550'W
Barnegat	184.4° T 197.5° M	10.8 kn	15° to starboard	35.055 nm	12:42 (3 hours 15 mins)	39°45.500'N 073°58.100'W
215 JACKS	200.0° T 212.9° M	10.8 kn	11° to starboard	106.759 nm	22:35 (9 hours 53 mins)	38°05.198'N 074°45.103'W
300A Chesapeake Light	212.0° T 223.9° M	10.0 kn	25° to port	83.710 nm	Feb 5 06:57 (8 hours 22 mins)	36°54.200'N 075°41.000'W
307 False Cape FALSE CAPE	186.1° T 197.1° M	10.0 kn	26° to port	18.807 nm	Feb 5 08:50 (1 hour 53 mins)	36°35.500'N 075°43.500'W

A few pages from a sample passage plan, including a blank spreadsheet showing a possible method of organizing information, a page showing waypoints, tidal graphs, and a printout of the weather forecast.

relevant information is gathered. The passage brief contains two types of information. First, there is information that is unrelated to the bird's-eye view afforded by the charts but which is vital to the plan—for example, weather, tides, currents, contact information. Second, there is information that reiterates, reinforces, or augments in greater detail the chart planning already completed. The spreadsheets above show different ways of organizing key information for each leg of the route, much of which also appears on the chart.

NEW YORK (The Battery)

Tuesday February 3, 2009
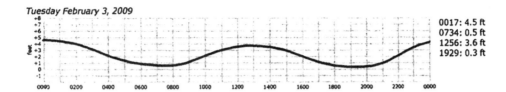
0017: 4.5 ft
0734: 0.5 ft
1256: 3.6 ft
1929: 0.3 ft

Wednesday February 4, 2009
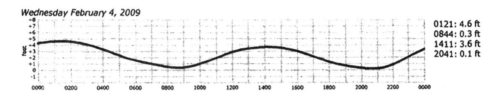
0121: 4.6 ft
0844: 0.3 ft
1411: 3.6 ft
2041: 0.1 ft

Thursday February 5, 2009
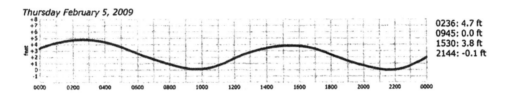
0236: 4.7 ft
0945: 0.0 ft
1530: 3.8 ft
2144: -0.1 ft

Friday February 6, 2009
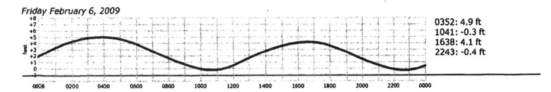
0352: 4.9 ft
1041: -0.3 ft
1638: 4.1 ft
2243: -0.4 ft

355 AM EST WED FEB 4 2009 .SYNOPSIS FOR THE COASTAL WATERS FROM SANDY HOOK NJ TO FENWICK ISLAND DE AND FOR DELAWARE BAY... LOW PRESSURE IS FORECAST TO PASS OFF CAPE HATTERAS THIS MORNING BEFORE MOVING DIRECTLY OUT TO SEA. MEANWHILE, HIGH PRESSURE WILL BEGIN TO BUILD IN FROM THE WEST FOR TONIGHT THROUGH SATURDAY. A WEAK FRONTAL BOUNDARY FROM THE WEST IS EXPECTED TO ARRIVE ON SUNDAY.

COASTAL WATERS FROM SANDY HOOK TO FENWICK ISLAND DE OUT 20 NM- SMALL CRAFT ADVISORY IN EFFECT THROUGH THURSDAY AFTERNOON

TODAY
N WINDS 10 TO 15 KT...INCREASING TO 15 TO 20 KT. SEAS 4 TO 6 FT.
TONIGHT
NW WINDS 20 TO 25 KT WITH GUSTS UP TO 30 KT. SEAS 4 TO 6 FT.
THU
NW WINDS 15 TO 20 KT WITH GUSTS UP TO 30 KT. SEAS 4 TO 6 FT.

COASTAL WATERS FORECAST NATIONAL WEATHER SERVICE WAKEFIELD VA

.SYNOPSIS FOR FENWICK ISLAND DE TO CURRITUCK BEACH LIGHT NC OUT 20 NAUTICAL MILES ...
LOW PRESSURE OVER NORTH CAROLINA WILL MOVE OFFSHORE THIS MORNING AND INTENSIFY AS IT MOVES NE INTO THE OPEN OCEAN. CANADIAN HIGH PRESSURE BUILDS INTO THE REGION TONIGHT THROUGH THURSDAY NIGHT. THIS HIGH THEN SETTLES INTO THE SOUTHEASTERN STATES THROUGH THE WEEKEND.

COASTAL WATERS FROM FENWICK ISLAND DE TO CHINCOTEAGUE VA OUT 20 NM- ...SMALL CRAFT ADVISORY IN EFFECT THROUGH THURSDAY AFTERNOON...

.TODAY...
N WINDS 10 TO 15 KT...INCREASING TO 15 TO 20 KT EARLY IN THE AFTERNOON. SEAS 5 TO 6 FT WITH A DOMINANT PERIOD OF 7 SECONDS. A SLIGHT CHANCE OF SNOW SHOWERS IN THE AFTERNOON.
.TONIGHT...
N WINDS 20 TO 25 KT WITH GUSTS TO AROUND 30 KT. SEAS 5 TO 6 FT WITH A DOMINANT PERIOD OF 7 SECONDS. A SLIGHT CHANCE OF SNOW SHOWERS.
.THU...
NW WINDS 15 TO 20 KT. GUSTS UP TO 30 KT IN THE MORNING. SEAS 4 TO 5 FT.
.THU NIGHT...
NW WINDS 10 TO 15 KT. SEAS 3 TO 4 FT.
.FRI...
W WINDS 5 TO 10 KT...BECOMING SW 10 TO 15 KT IN THE AFTERNOON. SEAS 3 TO 4 FT.

COASTAL WATERS FROM CHINCOTEAGUE TO CAPE CHARLES LIGHT VA OUT 20 NM- ...SMALL CRAFT ADVISORY IN EFFECT THROUGH THURSDAY AFTERNOON...

.TODAY...
N WINDS 15 TO 20 KT. GUSTS UP TO 30 KT LATE. SEAS 5 TO 6 FT WITH A DOMINANT PERIOD OF 8 SECONDS. A CHANCE OF FLURRIES IN THE MORNING...THEN A CHANCE OF SNOW SHOWERS IN THE AFTERNOON.
.TONIGHT...
N WINDS 20 TO 25 KT WITH GUSTS TO AROUND 30 KT. SEAS 5 TO 6 FT WITH A DOMINANT PERIOD OF 7 SECONDS. A SLIGHT CHANCE OF SNOW SHOWERS.

One challenge in creating a passage brief is determining what to include and what to leave out. The passage brief should not restate the contents of *Coast Pilots*, *Sailing Directions*, *Light Lists*, or other publications. The goal, in fact, is to distill that information to the bare essentials that are relevant to your vessel and your voyage. For instance, *Coast Pilots* contain a wealth of information on the biology and migratory patterns of the North Atlantic right whale, but all this information does not belong in a passage brief. You do, however, need to know your statutory obligations under the mandatory ship reporting system that protects these creatures if you want to avoid a hefty fine. Bullet points, subheads, and geographic or chronological groupings can be useful for organizing information in a passage brief. If necessary, you can reference the location where more information can be found and highlight it at the source.

Tides, Currents, River Stages, and Water Levels

Tides, currents, river stages, and water levels can have a significant impact on the plan, so this information should be addressed as appropriate. You first research tides and currents during the passage appraisal and planning, but in the execution stage you refine and calculate this information for the intended time of transit. In keeping with the concept of contingency planning, tide and current information should be available for a reasonable window of time on either side of the intended transit time in case the timing changes. River stages and water levels are influenced by many factors and cannot always be predicted well in advance like tides. If necessary, you can nail down this information in the execution stage and monitor it closely thereafter.

Weather, Ice, and Other Environmental Conditions

As we saw in the saga of the *Scandia*, the role of weather cannot be overemphasized in passage planning. Though good forecasting information was available, the captain did not integrate it into his planning or decision making in a meaningful way. The influence of weather on a passage plan depends entirely on the nature of the operation. In or near protected waters, a relatively short time horizon may be adequate for good decision making. For long coastwise or offshore passages, however, long-term weather developments may be more important to the plan. Seasonal and climatological variables such as hurricanes, ice, spring runoff, and wind patterns can affect planning in a big way, so it may be necessary to include information pertaining to these in the passage brief.

The significance of any particular weather concern depends largely on the capabilities of the vessel itself. In general, faster vessels have more control over where, when, and if they will encounter a storm system than slower vessels that can't get out of the way. In general, larger vessels are more robust and can take more punishment than smaller ones. That said, many a large, fast container ship has limped into port with cargo and lifeboats missing because of misplaced confidence in the ships' ability to outrun or out-tough a depression. Vessels operating on worldwide routes need to consider weather patterns not only at the point of departure and en route, but also at the destination. This kind of consideration drives the timing of yacht deliveries in many parts of the world.

The appraisal, planning, and execution stages each give rise to opportunities to consider the effects of daylight, darkness, and visibility on a given transit. If you find these factors to be significant, then your passage brief should address them.

Speed of Advance

Under the Rules of the Road, every vessel is expected to operate at a safe speed. This alone is enough to make speed of advance a factor for every watchkeeper, but the IMO's Guidelines for Voyage Planning further state that safe speeds should be considered during the planning stages. How fast is safe? Some argue that a safe speed is going as fast as possible without having an accident. It is not difficult to see problems with that definition.

Most vessels have an engine rpm or operating speed that represents a more or less optimal balance between fuel consumption and speed—a cruising speed, so to speak. Cruising speed is one obvious ingredient in the speed plan, and navigational and environmental constraints are others. In general, you expect higher speeds in open water and good visibility and slower speeds in more confined situations and restricted visibility. In waters where restricted visibility is frequent, the passage plan should provide for possible speed reductions in those conditions.

Sea state and operational particulars such as towing, fishing, or streaming research equipment also influence choices of speed. And then, of course, there's the schedule—the schedule, always the schedule.

In addition to these considerations, clearly there is a human element in determining safe speed, and the role of experience is not always instructive. Some inexperienced officers or officers on an unfamiliar route might prefer a slower speed so they can better maintain situational awareness. This is sensible. Other inexperienced individuals might underestimate the challenges of a particular leg, leading them to operate at a higher speed than they should. Oddly enough, seasoned mariners can fall on both sides of this issue, but for the exact opposite reason: experience has taught some to be cautious, while others who have successfully pushed the envelope in the past may become overconfident. For all these reasons the concept of safe speed from a planning standpoint may not be straightforward. But given that one purpose of passage planning is to minimize errors of judgment, the plan should provide explicit guidance regarding safe speed at any given point.

A speed plan is a good idea for other practical reasons, too. If the passage plan includes time-sensitive waypoints at which tide, current, or daylight is critical, monitoring the vessel's speed of advance is vital. Speed fluctuations are much more meaningful when measured against target speeds. While there are a number of legitimate reasons why actual speeds differ from planned speeds, there is no reason not to plan a target speed for each leg. As we have said, watchkeeping is largely about comparing what *is* happening to what *should be* happening, and speed is as valid a metric as any for this.

Run Times and ETAs

A run time is the duration of the voyage, or any leg of it, regardless of what the clock says. An ETA, on the other hand, is the estimated clock time of arrival at any point, including the ultimate destination. While ideally you hit all your ETAs as planned, in reality there are so many variables that even on a short trip you can't be surprised if you don't. Run times are based only on speed and distance, and therefore have a long shelf life so long as neither factor changes. In this respect run times are more versatile than ETAs, but ETAs remain an important tool for monitoring progress and are, of course, necessary when coordinating with external players such as VTS, port authorities, or pilots. Run times and ETAs are normally part of the passage brief, but some mariners put this information onto their charts for immediate reference.

Primary and Secondary Means of Position Fixing

IMO guidelines call for explicitly identifying the primary and secondary means of position fixing in a voyage plan. The choices here may seem obvious: in open water, use satellite systems backed by celestial navigation; in closer quarters, use satellite systems backed by radar. In many cases this approach is adequate. Whether to go further becomes a cost-benefit decision. Some will ask, with tools as accurate and reliable as these, why do more? It is hard to answer this question without appealing to professional pride. Since one of the goals of passage planning is to make sure you are doing things the *best* way possible, the sources of position data should be chosen to serve the situation, and this may mean looking past the most obvious tools.

There are at least three good reasons for prioritizing position-fixing methods. First, it forces deliberate consideration of what actually would be the best source of position information under specific circumstances. In many situations radar or visual methods trump GPS-based information, and we mustn't forget the utility of soundings. For instance, in regions where chart datum is unreliable, radar may be more reliable than latitude and longitude from a GPS. In other places, natural ranges and other charted features may be more useful indicators of the vessel's position. When this is the case, the passage plan should document these preferences for the benefit of all watch officers aboard.

Second, just as identifying primary sources is a focusing exercise, identifying secondary sources requires that you give some thought to what you will do if the primary source fails. This is a form of contingency planning that, sooner or later, proves justified.

Third, by identifying primary, secondary, and even tertiary sources, we put into practice the well-worn maxim to use "all available means." Since cross-referencing is the best defense against a navigational error, why not identify your options up front? The cruise ship *Royal Majesty* piled up on Nantucket Shoals in 1995 when a GPS cable chafed through. Successive watch officers didn't realize the receiver was providing dead-reckoned information rather than actual positions. Meanwhile, a perfectly functional loran unit sat nearby, but no one cross-referenced it with the GPS. Cross-referencing is not something you do only if you suspect something may be wrong; for a watchkeeper, cross-referencing should be what breathing is for everyone else—the moment you stop, you're dead. By identifying backup sources and putting them in writing, we are highlighting the very important role of cross-referencing on even the most modern bridge. If the captain of the *Safari Spirit* had embraced his navigational options more completely, the relatively undemanding transit to open water would not have cost him his vessel.

Fix Frequency

Most mariners appreciate that as navigational constraints increase, so should the frequency with which they confirm their vessel's position. On open ocean voyages with small chart scales, it may be acceptable to log positions hourly, plotting only once per watch. Nearer the coast, hourly plots may be necessary, and the interval will drop to half an hour and less inshore. When negotiating a truly tortuous stretch of water with few navaids and little room for error, position fixing is essentially continuous, sometimes requiring additional bridge personnel. By establishing a fix frequency, the passage plan can lift from any one individual the sole responsibility for initiating a tighter monitoring regimen.

A fix frequency for each leg of the passage plan sets forth a kind of schedule to follow, which is particularly helpful in preventing a team-of-one from fixating monitoring some other task to the neglect of position. The captain of the *Mercury* clearly needed this kind of prompting, and it was actually provided. If he had followed the stipulated fix frequency, he almost certainly would have avoided towing his barge over a ledge. A secondary benefit of an established fix frequency is that it clearly signals to a watchkeeper entering confined waters that a higher standard of care is now needed. Ideally, this heightened awareness carries over to other aspects of running the watch.

Also, an established fix frequency can help a watchkeeper monitor his or her own situational awareness. If the plan calls for cross-referencing your position every 15 minutes, and you find that you are not able to keep up, then you know that your situational awareness is slipping.

Perhaps the vessel is moving too fast or there is too much traffic to manage; perhaps distraction, stress, or fatigue is degrading your performance. Whatever the issue, position-fixing minimums provide benchmarks against which you can measure not only your position but your overall performance, too.

Finally, just as "safe" speed can be a matter of opinion, so too can adequate fix frequency. Establishing fix frequency beforehand helps prevent differences of opinion from producing different results with regard to passage monitoring. It harmonizes performance and reinforces the fact that each officer is a member of a team, even when alone in the wheelhouse. Fix frequency, like other aspects of passage planning, helps maintain the shared mental model.

Communication Plan

Imagine frantically paging through the *Coast Pilot* to find the VHF channel for hailing a bridge tender as the current is sweeping you down toward the bridge in low visibility and high winds. If the bridge tender is paying attention and you holler loud and long enough on Channel 16 and blow your horn, the bridge may open, but this is not the best approach, and there is no excuse for it.

Communications can be internal, meaning within the vessel, or external, meaning with everybody else. We have already noted the value of using chart notes to prompt both internal and external communications. In addition, the passage brief should contain anticipated contact information pertaining to the voyage. This could include phone numbers; VHF channels; email addresses for VTS; port, government, or terminal authorities; lock masters; bridge tenders; and pilots. If local conventions stipulate the use of sound, light, and flag signals, then this information should appear in the plan. If more detail is required, the passage brief should note where complete information can be found, rather than repeat a large amount of text in the brief.

*****

The purpose of passage planning is not to make work. The purpose is give the mariner an advantage in running a safe watch and making good decisions. The approach to passage planning set forth in this book is not nearly as onerous as it might appear. Much of this work has to be done anyway. You can either do it in advance or do it on the fly. Doing it in advance frees you to focus on monitoring and on making good decisions about truly unforeseeable events. It avoids the errors of judgment and calculation that often occur when you are plowing up a wake, juggling tasks, and trying to keep a grip on the bigger picture. Passage planning can be piecemeal or comprehensive. Rather than viewing the nuts and bolts of passage planning as a tangle of unconnected chores that must be fit in around other chores, a comprehensive approach encourages a more deliberate operating philosophy, which can only result in safer passages.

This discussion of passage planning tools and tactics is neither all-encompassing nor universally applicable. You may use some of these tactics and tools only sporadically, and others more routinely. Depending on the operation, a passage plan may need to address issues not even mentioned here. Still, opportunities to apply these ideas do arise, especially if you are on the lookout for them. Cataloging these tools and tactics here drives home the breadth of what is possible, and the absolute importance of taking control of the chart so that it serves your purposes.

For those whose sailing patterns are more itinerant the value of a detailed passage is more obvious. Those who operate only on familiar waters may have more to lose from complacency than a lack of information. Like other aspects of BRM, quality passage planning is more likely to occur when employers expect it and provide resources for accomplishing it.

Part II

Situational Awareness and Human Factors

SITUATIONAL AWARENESS is a watchword in the maritime industry and in any line of work where safety depends on people being highly attuned to what is happening around them. In the shipboard context a state of high situational awareness is characterized by the following:

- Accurately perceiving what is going on around you, both on and near the vessel
- Accurately discerning which developments have the potential to impact you
- Detecting *changes* to the present situation in a timely fashion
- Comprehending the significance of those changes for you
- Formulating a correct course of action for maintaining control of the situation

Situational awareness is a matter of degree, and it is inversely related to risk. The higher our situational awareness, the faster and more accurately we grasp what is happening, and the lower the exposure to risk. The lower our situational awareness, the slower and less accurate our responses, and the higher our exposure to risk. Ideally, situational awareness would always be at maximum, but it is just not possible. A newly minted mate reporting aboard a vessel for the first time cannot be as situationally aware as someone who has been working there for a time. A person who has been on duty for 15 or 16 hours cannot be as situationally aware as someone who is fresh. When you make the same trip on the same boat, doing the same job over and over, situational awareness can also decline, but for completely different reasons.

We can think of situational awareness as an agile defensive perimeter, a bubble, that is constantly expanding, contracting, and changing shape to meet evolving threats and circumstances. There are times for maximum alertness and times when lower levels will suffice. There are times to focus on the immediate task at hand (maneuvering, collision avoidance) and times to think ahead (passage planning, weather analysis). In a dynamic environment our perimeter must be highly responsive, yet this is asking a great deal, as there are many factors that undermine high situational awareness in the normal course of events.

While technology is useful in raising and expanding situational awareness by providing information, the human factors at the heart of the matter remain stubbornly fickle. All the technology in the world cannot prevent someone from growing complacent after performing a task for the umpteenth time. All the technology in the world cannot make someone alert if he or she hasn't had enough sleep. And all the technology in the world may be part of the problem when a solitary watchkeeper becomes distracted while steering through thick weather and calculating an ETA for VTS, monitoring the radar, chartplotter, AIS, multiple VHF channels, and answering a cell phone call from the boss, or home. Due to technology, we are able to accomplish more than ever, but we retain the same old standard-issue human faculties and frailties, therefore human factors continue to bedevil human performance.

It is one thing to talk about human factors abstractly, or even in the context of a casualty report. It is quite another to actually detect a loss of situational awareness as it happens to yourself or a shipmate. Unlike a piece of compromised equipment that emits an alarm to indicate a malfunction, or simply won't power up at all, people do not alarm when they begin to fail, nor do they have "acknowledge" buttons. Sometimes we have the presence of mind to cajole a little extra situational awareness out of ourselves and others ("Focus!"), but many of the human factors that degrade situational awareness are beyond our direct control, making them difficult to counteract in the moment. If we are to manage the human factors that impact situational awareness, we must first become acquainted with them and then use that knowledge to shape operational choices. By probing the interaction between situational awareness and human factors, we can begin to see that these often elusive issues are real, not imaginary, and not peripheral to a safe operation. In following chapters we examine some well-documented human factors that are known to affect situational awareness, including the following.

Overreliance: Technology has the power to shape attitudes toward risk. Increasingly accurate, reliable, and sophisticated electronic navigational equipment can adversely affect a watchkeeper's perception of risk when it promotes the uncritical acceptance of crucial information. The human-equipment interface is an area that can both raise *and* lower situational awareness, depending on how we approach it.

Distraction: Job related or not, distraction is a fact of life aboard. Many of the most pervasive shipboard distractions are in fact related to getting the job done, it's just that they come at the wrong time. This fact sometimes makes it difficult to discern between a timely interruption and an irrelevant distraction.

Stress: No job is stress free, but excessive stress can muddle priorities as mariners try to reconcile job-related pressures with safety considerations.

Fatigue: One of the challenges of an intense, round-the-clock work environment is obtaining adequate rest. The issues surrounding fatigue are well documented and should be understood at the operational level as well as the management level.

Complacency: Of all the human factors that degrade situational awareness, complacency is probably the most unavoidable. This, however, should not discourage mariners from watching for it and attempting to manage it.

Transition: Transitions are created by circumstances rather than human factors, but they represent points of vulnerability in human systems (i.e., a ship's crew). Transitions can create stress and distraction, exacerbate fatigue, and exploit complacency. Life underway is full of transitions, and to the extent we are alert to them we can anticipate and manage their affect on situational awareness.

The issues described above are concerns for people everywhere. What is different for mariners is that, due to the nature of the work, there is potential to do great harm on a large scale. Plans and procedures are fine things when we can anticipate a role for them, but it is situational awareness that keeps us on track after the pavement runs out. The incidents and the themes discussed in this section can serve to warn us of how situational awareness slips away. And remember, we are concerned not only with our own situational awareness but also with that of the people around us.

Chapter Five

Overreliance

OVERRELIANCE AND COMPLACENCY are close cousins. The former amounts to putting too many eggs in one basket, and the latter leads us to not watch the basket closely enough. Although everyone relies on *something* to get the job done—no one is completely self-sufficient—when something goes terribly wrong, ordinary reliance can suddenly be revealed as overreliance. The transition from reliance to overreliance can be subtle, and usually takes place over a period of time. Consequently, the accompanying loss of situational awareness is difficult to detect.

Overreliance can pertain to all of our resources: information, equipment, and people. But the focus here is on the interface between bridge equipment and watchkeeper, and the nature of that relationship. Mariners of old, who relied on dead reckoning, bottom samples from a lead line, and some unquantifiable "feel" (instinct shaped by experience) for what was happening in fog or darkness, lived in constant doubt. There was little in their daily experiences to encourage overreliance on any one source of information. Old-fashioned methods involved considerable finesse and caution. Standing off until dawn, having an anchor ready, waiting for fog to lift, and checking to see if the taffrail log was fouled with seaweed and therefore underrecording speed were routine defenses against disaster. Navigators were conservative, as there was no intelligent basis for being otherwise.

Mariners of today do not live with such doubts. The equipment they use is (usually) accurate, reliable, and much easier to use. Daily experience encourages mariners to place their faith in the equipment because it is, in fact, very good. Often it is more accurate than the mariners themselves; a GPS gives a much more accurate position than a star fix. So you can hardly blame navigators for operating in a manner that might have been considered imprudent by past standards. But when reliance on these modern, accurate pieces of equipment turns into overreliance, and that overreliance leads to an incident, the problems with uncritical acceptance of such data become glaringly obvious.

Despite all this technology, even the most gadget-happy young salt knows deep down inside that placing total faith in any single navigational system is folly. But it is so easy to come to rely on a piece of equipment that has "proved" itself in the past. The only real defense against overreliance is cross-referencing electronic data with other sources of information and comparing the results. But how far should cross-referencing go? At what point has due diligence been exercised?

There are no simple answers, but we do know that fortune favors the prepared mind, and some people are more likely than others to commit the sin of overreliance. People who catch a mistake or problem in the nick of time may appear to be beneficiaries of good luck when, in fact, an underlying mental preparedness led them to notice what others did not. We also know that some circumstances are more likely than others to breed overreliance in people who normally would not succumb. Factors such as stress, distraction, fatigue, and complacency can all induce moments of misplaced trust. The golden cautionary rule with regard to overreliance is that just because everything seems fine doesn't mean it is fine. The "magic box" that suddenly starts producing incorrect data does not play favorites; individuals who are aware that magic boxes can spit out erroneous readings are better prepared to catch the error and respond appropriately.

The *True North*, 2004

The *True North* was a modern 113-foot aluminum-hulled vessel specializing in adventure tourism in the remote Kimberley region of Australia's northwest coast. The vessel was equipped with twin screws, a bow thruster, five motorized launches, and a helicopter for touring the remote wilderness territory. The bridge console was arranged for easy monitoring and

The *True North* was a fully equipped, modern vessel serving Australia's northwest coast. (Australian Transportation Safety Bureau)

cross-referencing. It included radar, a color depth-sounder, forward-scanning sonar with a color display, two GPS units with "smart" antennas, two personal computers running separate electronic chart systems (ECS), and an autopilot. The bridge also had a large and very comfortable chair for the watchkeeper.

There were three main modes for controlling navigation aboard the *True North*. In the first mode, autonav, a GPS unit fed position data into one of the two electronic chart systems, which in turn fed information to the autopilot. (The ECS in use was a commercial product but was not an IMO-compliant ECDIS. This was not required because the vessel was not subject to the provisions of SOLAS.) The autopilot made the necessary course adjustments to keep the vessel on the programmed route. In the second mode, the watchkeeper would disengage the ECS and enter the desired course directly into the autopilot. This mode was primarily used for large course alterations that the watchkeeper would make in increments. The third mode was hand steering, which was used only for maneuvers such as docking or

The *True North*'s bridge console was equipped with radar, personal computers running chart software, and an autopilot arranged for easy cross-referencing. The wheel is just to starboard. (Australian Transportation Safety Bureau)

The *True North*'s console and wheel, showing the location of the GPS cross-track error screen as well as the magnetic compass. (Australian Transportation Safety Bureau)

undocking. Autonav was by the far the most used mode and was considered reliable.

The *True North*'s captain was thirty-one years old and had been working on vessels in Western Australia for fifteen years. He had worked extensively on the Kimberley coast as a mate and a captain on passenger vessels, including a hitch as mate on the *True North* the year before. Aboard the *True North*, watchkeeping duties were customarily shared between the captain and a mate.

On August 7, 2004, the *True North* was halfway through a two-week cruise with twenty-six passengers and twelve crew on board. The immediate destination was the St. George Basin in the Prince Regent River. The basin and much of the surrounding coast are known for 30-foot tides and strong currents. Due to the remoteness of the region and the lack of commercial traffic, much of the coast is unsurveyed and there are few navaids (lights, buoys, etc.). However, the captain was familiar with the route: Sailing as mate the year before, he made at least seven trips into the St. George Basin. Earlier in 2004 he had been promoted to junior captain and had made three more round-trips to the basin under the supervision of the company's senior captain. He had since made two additional transits on his own. On each occasion, the vessel had been navigated using the autonav mode.

The workday began at 0600. The vessel got underway from the previous night's anchorage and shortly after midday anchored in Porosous Creek en route to the St. George Basin. The passengers went ashore to explore while the captain took a launch to do some fishing for an evening barbecue. At 1600, with the passengers back aboard, the mate weighed anchor and took the vessel back down Porosous Creek, where he picked up the captain. The vessel anchored again near a small island and the passengers went ashore for a beach barbecue. Afterward, the *True North* got underway one more time for the trip into the

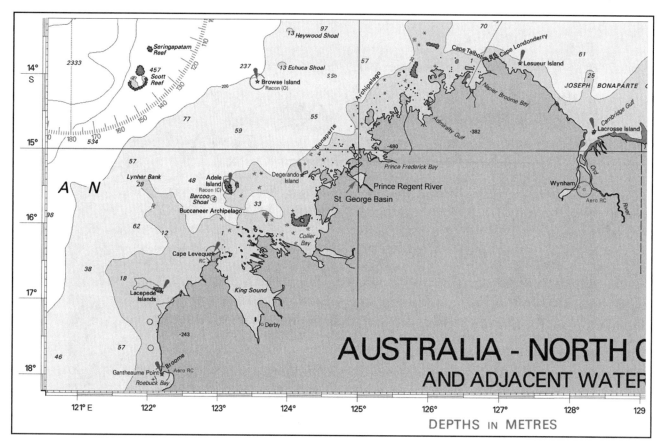

The *True North*'s accident occurred in the St. George Basin on the Prince Regent River.
(Australian Transportation Safety Bureau)

St. George Basin. After socializing with the passengers, the captain retired to his cabin for about an hour while the mate stood watch. At 2200, with the vessel approaching narrower waters, the captain returned to the bridge and took the conn. He had been on duty in one form or another for 15 hours, with a 1-hour break before going on watch. This work routine was not uncommon during the vessel's cruises in the Kimberley region.

The vessel was operating in autonav mode, making 11 knots on the same preprogrammed route that had been used in the past. The radar was on but the depthsounder and the forward-scanning sonar were not. In order for the forward-scanning sonar to operate, the transducer had to be lowered beneath the hull. Since the transducer had been damaged more than once by debris, it was no longer used on preprogrammed routes as they had no known dangers. Regular checks of the ECS showed that the vessel was right where it was supposed to be on the computer-generated track line. As the tide began to flood, the *True North*'s speed over the bottom increased to 12 and then 13 knots.

Upon approaching the narrows at Strong Tide Point, the captain stepped out of the wheelhouse to have a look around. Suddenly aware of the proximity of shore, he hurried back to look at the radar and saw that the *True North* was heading straight for the point at 14 knots. The captain pulled back on the throttles just as he heard a loud crunching sound. At 2304 the vessel came to a stop.

The crew was able to contain the flooding and refloat the vessel on the rising tide and anchor it in deeper water. The next day the captain beached the boat and surveyed the damage at low tide. The passengers were evacuated, and a damage control team was flown in by seaplane. The team spent the next two days making temporary repairs, after which the vessel limped back to Darwin, where it spent eleven days in a shipyard undergoing permanent repairs.

The captain told investigators later that he had been cross-referencing by radar, yet his vessel grounded 1,000 feet to the north of where it was supposed to be. Running aground at a place called Strong Tide Point might suggest that tidal currents were involved.

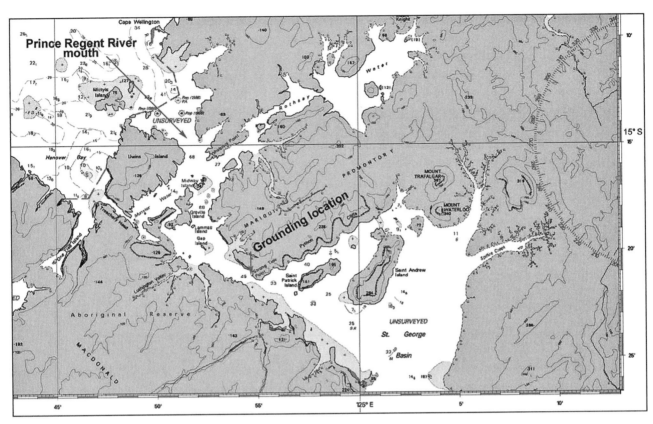

A portion of chart AUS 730 showing St. George Basin and approach passage. (Australian Transportation Safety Bureau)

However, the autopilot was capable of handling set and drift, and a replay of the vessel's course over the ground showed that there had been no unintended course alteration in the minutes before the vessel ran aground. The main thing was this: The actual position where the *True North* grounded differed significantly from where the ECS showed the vessel to be. Even as the vessel was stuck fast to the bottom, the electronic chart showed it in good water, right on the electronic route, 1,000 feet away. Something didn't make sense.

Failure to Cross-Reference

Several questionable watchkeeping decisions were exposed in the review of this incident. For instance, why was the captain so pressed to be in the St. George Basin in the middle of the night that he was making 14 knots in darkness through a winding, narrow, and poorly surveyed channel? It was undoubtedly related to the schedule and catching a fair tide, but this was a pleasure cruise, not an emergency, and there was another fair tide in the morning. And why, in those unsurveyed waters, was the forward-scanning sonar turned off, the one thing that could have alerted the captain to shoal water ahead? Unfortunately, protecting the transducer had become a higher priority than protecting the boat. There was a logic to it, but in this case it proved to be flawed. Depth information was critical but it was not available because it was not standard practice to integrate information sources on the *True North* in this way. Most important, however, was the captain's overreliance on the ECS.

Despite his claims to the contrary, the captain did not adequately cross-reference the electronic chart data. The vessel was 1,000 feet from where it should have been at the narrowest part of the transit, with a fully functional radar. The approach was particularly well suited to parallel indexing, and indeed, the company's senior captain had directed watchkeepers to use radar as the primary position-fixing tool in that area. However, these instructions were not formalized into the vessel's SOPs or the passage plan. Investigators concluded that, regardless of what the ECS indicated, if the radar had been closely monitored, it would have been obvious that the vessel was standing into danger.

For any cross-referencing to be effective, it must keep pace with the speed of advance and the inherent challenges of a route. It is self-evident that the captain was not able to keep up. He had had a long day and he did not enlist the support of the mate. No target speed had been established in the passage plan, so he was literally going with the flow and ended up leaning heavily on the ECS. The system had earned his trust and he did not anticipate the potential for error. Since the ECS had worked in the past for the same voyage, its performance this trip raised an important question: what had changed?

Investigators were able to re-create the *True North*'s trip from information stored in the ECS. Upon review, it appeared that a position anomaly occurred while the vessel was anchored at Porosous Creek on the afternoon before the grounding. The effect was to shift the vessel's electronic position almost 700 feet to the west, high and dry. As the ECS record shows, the vessel's inbound positions closely followed the programmed route but the outbound route shows the vessel transiting the western shore, well outside the creek bed. This deviation did not catch the attention of the mate, who simply retraced the vessel's path out of the creek visually without closely examining or questioning the electronic chart. Once the *True North* was in more open water, the position discrepancy was not so obvious, and the anomaly had no consequences until the vessel approached Strong Tide Point. By this time darkness had fallen, making it more difficult to

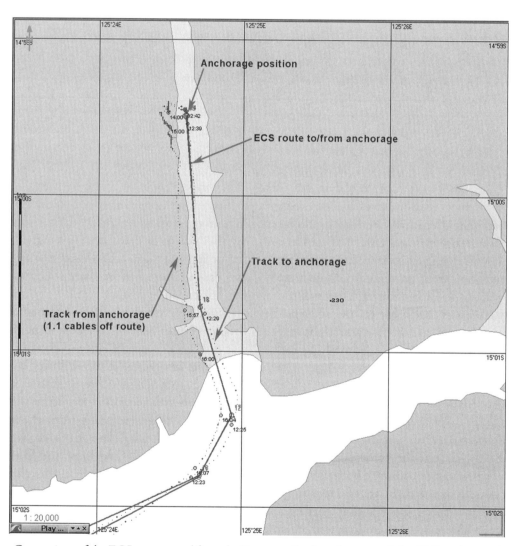

Comparison of the ECS route to and from the Porosous Creek anchorage and the GPS track to and from the same anchorage. The mate did not catch the discrepancy between electronic display and the actual course made good when it was easiest to detect. (Australian Transportation Safety Bureau)

detect the problem by eye. At both Porosous Creek and at the time of the grounding, the ECS consistently placed the vessel to the right of the actual track made good.

Ultimately, the investigators uncovered several scenarios that could have led to the ECS displaying erroneous information. All the scenarios are extremely instructive to anyone using electronic charts aboard small commercial vessels. And yet, from a watchkeeping standpoint, does the technical explanation for false information in one piece of bridge equipment really matter when we know that the possibility of false information applies to *all* of our bridge equipment, *all* of the time? The resources were available to catch the error but they were not effectively managed because the captain had developed a tendency to uncritically accept what the ECS showed him.

The investigators first looked at the two personal computers that ran the *True North*'s ECS programs. They learned that these computers also ran numerous other programs and applications relating to provisions, catering, maintenance, and other operational needs. This is not uncommon; on many smaller vessels, computers run navigation programs as well as programs for email, tides and currents, weather, bookkeeping, spare-parts inventories, and so on. The investigators said the following about this practice:

> ECS are susceptible to changes in computer operating system files which occur when other programs are installed or deleted from the personal computer's hard drive. Changes . . . may alter the operation of the electronic charting program leading to errors and its incorrect operation. The most common problems come from computers that are used for more than one task and have many programs loaded on their hard drives. This results in conflicts when the user attempts to multitask between programs while running ECS in the background.

The investigators were unable to prove conclusively that multitasking of the ECS computers caused the position anomaly, but the existing arrangement made it possible, and there were no safeguards in place to prevent it. They strongly recommended that computers used to run navigational programs be dedicated solely to that purpose.

Second, the investigators examined the geodetic datum used aboard the *True North*. Several geodetic datums are used in Australia, and more than one is used for the Prince Regent River area. Two of these,

AGD (Australian Geodetic Datum) 84 and AGD 66 were analyzed and compared with the WGS 84 datum that GPS typically utilizes.

The results showed that AGD 84 and AGD 66 positions plotted almost 700 feet south and west of WGS 84 positions on a WGS 84 chart. In this instance, if a mismatch occurred between the chart datum and the electronic chart in use, it would show the vessel about 700 feet to the right of the track *True North* was actually following. This is about the same offset that the ECS recorded when the vessel departed Porosous Creek. If the datum differential were applied to the ECS positions at Porosous Creek, it would place the vessel back in the center of the creek, on the programmed route. Investigators were unable to prove that the wrong geodetic datum was in use at the time of the grounding; however, an inadvertent datum switch was possible, and many GPS users are wholly unaware of this issue.

Third, the investigators looked at the GPS system that fed position data to the *True North*'s ECS. It was fully functional at the time of the grounding and there were seven satellites in view of the vessel's GPS antennas, with sufficient geometry to produce good data. However, three noteworthy points emerged from the investigation regarding the quality of the GPS data.

The first point was that at the time of the grounding, a substantial portion of the sky, 31%, was not visible to the GPS antennas. This was due to the combined effects of the antenna height (only 26 feet above the water) and the close proximity of land features, which completely surrounded the vessel. Where land features are higher, more of the sky is blocked out. A restricted view of the sky is an operational reality in some places (in narrow fjords, near tall buildings). Smaller vessels with lower antennas are more vulnerable to reception problems than vessels with higher antenna mounts. Under some circumstances, GPS data may suddenly be interrupted, even though the GPS system is fully functional. When this happens, the GPS unit *should* emit an alarm, but what if it doesn't?

The second point was that during repairs to the vessel a technician found that several of the GPS antenna connections were loose at the terminals in the wheelhouse. Obviously, loose cabling can lead to intermittent signals, which can corrupt any navigational system.

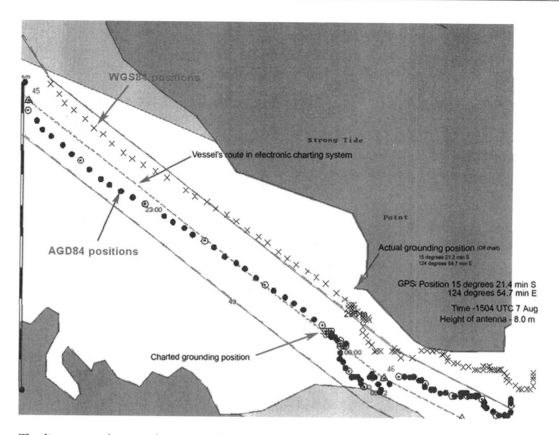

The discrepancies between the *True North*'s route using the WGS 84 datum and the AGD 84 datum. There may have been a mismatch between the datum used by the GPS unit and the chart in use. (Australian Transportation Safety Bureau)

The third point related to GPS data was that investigators noted that on rare occasions the GPS system as a whole is subject to certain inherent errors, known as "blunders," which can result in large position errors. A very large blunder would presumably be obvious to a navigator, but a smaller yet still significant error might not be.

The prevailing opinion was that the position discrepancy experienced aboard the *True North* did not originate with the GPS data, but the issues mentioned above stand as reminders that there are many ways GPS data can be interrupted or degraded. How an interrupted or degraded GPS signal manifests itself to a watchkeeper varies with the equipment and may not always be obvious. Cross-referencing with other sources may provide the first inkling that something is not right.

Finally, the investigators noted that the ECS aboard the *True North* was not running official government

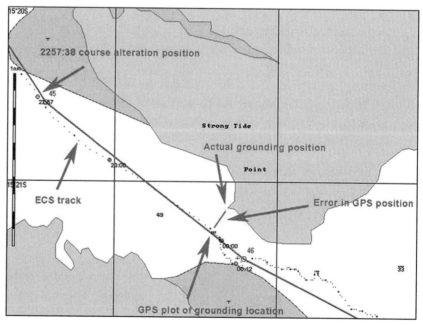

The ECS plot of the *True North*'s track leading up to its grounding. (Australian Transportation Safety Bureau)

charts. It was using electronic raster charts, digitized reproductions of paper charts published by the Australian government. Although the raster charts were accurate for plotting WGS 84 positions, investigators noted that the accuracy of the position of land features depicted on them could not be determined in all cases. This very issue was raised in the discussion of passage planning (Chapter 4). The region in which the *True North* operated contained large areas that were minimally surveyed or not surveyed at all, and there were no navaids on this particular route. When this is the case a prudent navigator takes extra measures to avoid the sort of overreliance that led to this incident.

Gadgets and Gizmos

Overreliance on gadgetry in all its forms seems to be a widely shared human weakness, making cautionary tales like that of the *True North* useful but hardly unique. Setting aside the issue of erroneous positions being caused by deliberate satellite jamming, many mariners can attest to having witnessed peculiarities, erratic performance, and downright wrong information coming from their electronics. The core navigational curriculum at the school of hard knocks regularly reminds mariners that the word "autopilot" has always been an oxymoron. The tendency to embrace ARPA-generated data as gospel has educated countless mariners who accepted small CPAs and high speeds only to learn the limitations of their equipment disturbingly close to another vessel. What is sobering about the *True North* case is that investigators were unable to pin down the cause of the ECS anomaly precisely because of the multitude of reasonable explanations. If we extrapolate these explanations, and all other potential explanations for electronic glitches across the industry worldwide, it is clear that overreliance in the human-equipment interface carries genuine risk.

Bridge equipment in all its forms constitutes a major portion of our bridge resources. Just as situational awareness is a flexible perimeter that is always adjusting to the situation, mariners should always be adjusting their reliance on bridge resources to the situation. No one piece of equipment does everything, though sometimes mariners treat favored tools as if they can. The secret to the effective management of bridge equipment lies in seeing the individual capabilities as complementary, using all resources judiciously, and overrelying on none. On a quiet, starlit night with plenty of sea room and no traffic, a more relaxed approach to position monitoring might be justifiable. But barreling through narrow, uncharted waters in the dark calls for a higher standard of care.

*****

Some argue that removing much of the doubt from navigation with modern systems has "dumbed down" the traditional challenges of watchkeeping, leaving mariners bored, complacent, and ripe for the kind of overreliance that led to the grounding of the *True North*. Autonav systems are only one example of a technological innovation that warrants a health warning. Some mariners maintain that such systems disengage the watch officer from the job of watching, and there is almost certainly some truth in this. While these devices free up watchkeepers to focus on other things, they can also free up people to tune out, and therein lies a hazard. Today a navigator's energy goes less into determining the vessel's position and more into confirming it by checking, double-checking, and triple-checking. If done right, however, this too is a demanding job that requires an active mind and critical thinking. The skepticism regarding autonav systems is perhaps no different from the skepticism heaped on earlier technological advances, such as when radar was first introduced on ships. But it may also be true that the effect of seeing your vessel displayed in miniature on an electronic chart is particularly seductive. They say seeing is believing, and most of us do want to believe.

Reliability is a valued virtue that can usually be found in the equipment (and the people) we sail with. Problems arise, however, when a device that has always been reliable suddenly fails. This usually comes as a surprise, but it shouldn't. To err is human, and we should certainly expect it of machines. Passage planning and procedures that institutionalize using multiple sources of information can help prevent the gentle slide from reliance to overreliance. Training watchkeepers in the capabilities and limitations of specific equipment is another important defense. But the best defense against overreliance is training watchkeepers to cross-reference frequently, which includes applying traditional methods.

Chapter Six

Distraction

Distractions are a normal part of a complex life, and running a vessel is certainly a complex task. Sometimes distractions are welcome when they provide a stimulating alternative to the task at hand. There may be a time for that. On a boat there are plenty of distractions and mariners must learn to see them for what they are and manage them. Succumbing to distraction is not a crime—unless, of course, it leads to something serious.

The *Queen of the North*, 2006

The *Queen of the North* was a roll-on, roll-off ferry certified for 650 passengers that ran a regular route along the Inside Passage of British Columbia. Strictly speaking, it was not a limited-tonnage vessel, yet its tale is instructive to the small-vessel world. The watch rotation was comprised of alternating 12-hour shifts, each with two licensed officers and four deckhands (also known as quartermasters). The *Queen of the North* was customarily operated with a bridge team, as opposed to a lone watchkeeper, although the precise configuration of the team changed with the circumstances.

On the night of March 21, 2006, the *Queen of the North* was making its first regular run after a four-month refit. As part of the refit, a new steering-mode selector switch had been installed. Although the procedures for the new switch had been posted on the bridge, not all of the deck crew were familiar with it yet. The second and fourth mates had the 1800 to 0600 bridge watch. It was common practice for the two officers to relieve each other throughout the watch, and then double up for areas deemed to require more situational awareness. A little before midnight, as the vessel approached the relatively open waters of Wright Sound, the fourth officer relieved the second officer so that he could take a break. The fourth officer had been working for the ferry service for fourteen years, eight of them as an officer; the second officer had twenty-five years with the ferries. Before leaving for his break,

The *Queen of the North* underway in British Columbia. (Kevin Stapleton)

the second officer set up a laptop computer on the bridge to play some music.

Also on the bridge was a deckhand. Since the vessel was on autopilot at the time, the deckhand's primary responsibility was looking out. At 2359 a minor course adjustment was made to maintain the track line. After checking the proximity to the next VTS call-in point, the fourth mate dimmed the ECS to preserve his night vision. The electronic chart in use at the time was a raster chart, which was brighter than the vector charts aboard. As a result, some officers had formed the practice of dimming the screen at night except when they wanted to check their position.

At 2 minutes past midnight, now March 22, the fourth mate checked in with the VTS dispatcher. The vessel was making 17.5 knots and was about 5 minutes from the next course change. A squall went through, briefly reducing visibility before clearing off. A few minutes later the *Queen of the North* passed the waypoint without making the planned course change.

At 0020, 14 minutes and 4 miles after missing the course change, the fourth mate abruptly realized the error and ordered the deckhand to alter course to port. As the deckhand went to make the course change, she noticed trees going by the window. A passenger later reported having watched land go by for several minutes. The fourth mate ordered the deckhand to switch from autopilot to hand steering but the deckhand was one of the crew who did not fully understand the new steering system. Seconds later the vessel struck Gil Island at full speed. The ship continued forward for some time, scraping and banging off the rocks, before drifting away from the island severely holed.

Though the ferry remained afloat for 77 minutes, the forty-two crew members were unable to fully account for the fifty-nine passengers aboard. The vessel lacked an evacuation plan, and the clearing of the interior spaces was by turns haphazard, incomplete, and redundant. In some compartments, the icy water was waist deep before the last person was out. When the ship finally slipped under, two passengers went with it and were never found. Given the multitude of problems encountered during the emergency response, it is indeed fortunate that there were only fifty-nine passengers aboard, not the 650 for which the vessel was certified.

Failure to Pay Attention

You don't ever want to be the subject of an investigation. Aside from the obvious reason—that something must have gone horribly wrong to warrant the attention in the first place—investigations uncover all kinds of disturbing issues that seldom reflect well on anyone. Though our interest in the *Queen of the North* centers primarily on the issue of distraction, this case

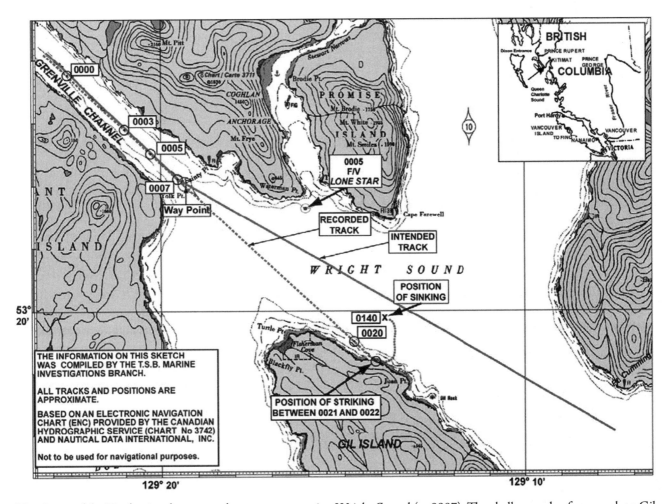

The *Queen of the North* missed a course change upon entering Wright Sound (at 0007). The shallow angle of approach to Gil Island left ample time to discover the error had the radar or ECS been monitored. (Transportation Safety Board of Canada)

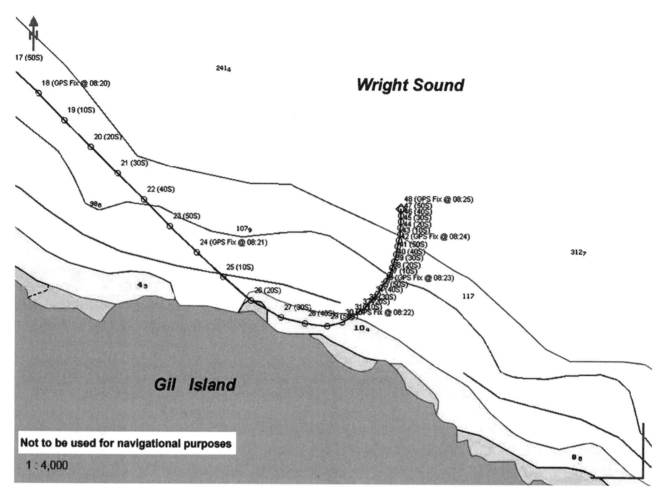

The electronic chart system's recording of the *Queen of the North*'s track. After it struck Gil Island at full speed, the ferry carried on for several more minutes. Two passengers perished. (Transportation Safety Board of Canada)

is an example of how an outwardly professional operation can harbor a multitude of other problems.

As it turned out, the fourth officer and the deckhand who were on watch during the incident knew each other quite well. They had been in a personal relationship that had ended two weeks earlier. This watch was their first encounter without other people around since that time. Sometime in the minutes after the VTS call-in, they engaged in what investigators described as "a personal conversation" unrelated to the safe conduct of the vessel. The nature of that conversation has, perhaps understandably, generated a lot of speculation. And yet, like pinpointing the precise cause of the ENC malfunction discussed in the last chapter, from a watchkeeping standpoint it hardly matters. Whatever transpired, the bottom line is that these two people were not overwhelmed by circumstances. They simply ceased to pay attention to where the ship was going.

The fourth officer explained that he thought he *had* ordered the planned course change right after speaking with VTS. While this could have been an attempt to deflect blame, anyone who has committed an oversight due to distraction knows that his response is consistent with the way distractions often work: A person may be conscious of what needs to be done and may even take steps toward doing it, but then becomes sidetracked by something untimely or irrelevant and moves on without completing the task. Later, when the mistake is revealed, the tendency is to say, "I thought I did that," or "I thought you did that." In this case the mate resumed his interaction with the deckhand without making the course change.

In addition to the issues mentioned above, investigators uncovered several safety concerns. First, Canadian regulations called for a third qualified person to be on the bridge. If someone else had been on the bridge, it is more likely that the missed course change would have been noticed, or that someone might actually have looked out and seen the island while there was still time to react. It is also conceivable that a more professional atmosphere would have prevailed on the bridge with a third person present.

Second, the ECS, a significant bridge resource, was dimmed in order to preserve night vision. Although this was common practice aboard, it deprived the mate of the real-time, monitor-at-a-glance benefits that electronic charts are supposed to provide. A casual glance at any of the three radars should have revealed that the vessel was standing into danger; nevertheless, allowing watchkeepers to choose between maintaining night vision and monitoring the ECS was a bridge resource management flaw that was bound to reveal itself in a moment of low situational awareness.

Third, alarms on the ECS and radars relating to cross-track error or the approach of danger targets, waypoints, and course changes were turned off, unavailable, or reconfigured so that they were inaudible. Ironically, the crew considered the alarms distractions, as have many people.

Fourth, when investigators interviewed the crew, they received four different explanations for how the new steering-mode selector system worked. The steering system and the helmsman are significant bridge resources. They control where a vessel goes. A helmsman who doesn't understand the steering system is an obvious liability to a bridge team. While it is debatable whether a quicker switch to hand steering would have saved the ship, the incident highlights the importance of truly understanding the details and the nuances of your resources when taking charge of the watch. It also highlights how changes to familiar resources can leave mariners vulnerable to mistakes under pressure. To preempt this type of breakdown, some sort of training investment is necessary before a critical situation arises.

Fifth, against regulations, watertight doors were in the open position at the time of grounding, which contributed to continuous progressive flooding. The purpose of watertight subdivision is well known. Leaving them open underway is risk-taking.

Sixth, life rafts were not fitted with float-free releases, nor were they required to be. Sometimes regulations lag behind common sense, and when they do, responsible people do the obvious thing without waiting to be told. Far too much is known about this particular issue to excuse the suspension of common sense.

Finally, investigators learned that, despite a zero-tolerance policy, some crew members in safety-critical positions used marijuana both aboard and in port. No information emerged linking the two watchstanders to this practice, but since no post-accident toxicological tests were performed, the issue could not be laid to rest.

Sex and Drugs and Rock & Roll

Predictably, the media had a field day. Shipboard romance, music playing in the wheelhouse, and dope smoking by the crew are among the details that will forever surround this case. Regardless of the media frenzy, the point is that this event was a fatal tragedy for some and a professional and personal catastrophe for others. Whatever the other operational shortcomings aboard the *Queen of the North*, it was the inability to separate the private from the professional that sank a ship and killed two people. If the two people on watch had paid just a little more attention, all the other issues would have been nonfactors, at least for the time being.

In most cases, the argument favoring a bridge team is that it allows the workload to be distributed so that the person in charge can preserve overall situational awareness and prioritize effectively. That argument remains valid, yet this case illustrates a stark example of a bridge team–assisted casualty that was a direct result of distraction.

Because people are social creatures, virtually all relationships and interactions have the potential to distract, not only those with a titillating, romantic backdrop. A captain entering a wheelhouse can distract an officer who is overly deferential and tacitly relinquishes control, or who gets caught up in the old man's sea stories and stops looking around. Shipmates can distract one another through casual conversations that help pass the time but draw attention away from the job at hand. Smoldering animosities between individuals, entire shipboard departments, or between the crew and shoreside management also have the power to distract. Passengers or guests are definitely distractions, especially if they are allowed around the helm station and if hospitality is part of the crew's job description. Pilots, shoreside bosses, and trainees all can be distractions if crew allow themselves to become overly invested in their presence and lose track of their primary duties. Nor is the circumstance at play on *Queen of the North*—men and women working together—so rare a shipboard phenomenon these days that we can afford to dismiss it as exceptional.

The fourth mate and the deckhand must have been chagrined when, during the post-accident investigation, music could be plainly heard in the background on the recordings of their radio exchanges with VTS. Combining watchkeeping with any kind of entertainment may be inconceivable to some people, yet others vigorously defend the practice of playing music on watch, insisting that it helps them stay awake and engaged, particularly in open water with no traffic. It would be difficult to prove them wrong with one incident. But even if true, we must still recognize the potential for such entertainment to distract and degrade situational awareness. Anyone who has ever unconsciously tapped a foot to a musical beat knows that while that foot is tapping, some part of the brain is not on the job. Where do you draw the line with this, and who draws it? Music, radio, audiobooks, and similar forms of entertainment not only have potential to distract watchkeepers but they also reduce situational awareness by masking critical sounds such as signals or alarms, mechanical problems, or someone calling for help. The fourth mate and the deckhand on the *Queen of the North* can declare until their dying day that the music on the bridge had nothing to do with their loss of concentration, yet to anyone else listening to these tapes the impression is deeply unprofessional in light of what occurred.

Entertainment on watch makes a statement about how to approach the job of watchkeeping, and it can shape behavior at all levels of an organization. If the mate can listen to music on watch, why can't the lookout? And to anyone aboard another vessel, music in the background of a radio transmission sounds like a party in progress. This erodes bridge-to-bridge trust and can only complicate the interaction.

Too Many Distractions

Multitasking has become the word of choice to describe any situation where people are trying to do a number of things at once. The term is often used to connote pride in our ability to keep a situation under control when fielding a variety of complex demands. However, the notion that multitasking is the key to productivity can be an illusion. Research shows that, in reality, people cannot do multiple things at once, or at least not do them well. In his book, *The Myth of Multitasking*, David Crenshaw explains that what people really do is *switchtasking*—rapidly shifting focus from one thing to another. Switchtasking doesn't necessarily translate into increased productivity or cognition; in fact, the opposite may be true. Common sense should tell us this, because most of us already know that when we try to do too many things at once we end up making mistakes, some of them costly. Yet for some reason, people often seem to find multitasking fulfilling. Switchtasking, a more accurate term, can be a useful skill that is sometimes necessary, but it comes with costs that should be understood.

Switchtasking involves transitions, and all transitions have costs. Each time you switch focus from one task to another, you lose concentration and fluency. You have to drop the mental tools for one project, then find, pick up, and begin to use the appropriate tools for the next. Each time you do this there is a transaction cost, a slight loss of ground or momentum that you have to recover before the new activity becomes productive. By some estimates, completing a task can take twice as long when switchtasking. When you switch back to the first task or move onto a third one, these transaction costs are incurred again. The costs can accumulate, muddling priorities and lowering overall situational awareness. When switchtasking reaches a certain level of intensity, you may be accomplishing very little despite the appearance of activity.

Not only is switchtasking inefficient, but it also creates fertile ground for distractions. One study of long-haul truckers found that the risk of collision is twenty-three times higher when drivers are text messaging. Big surprise. This has obvious ramifications for marine operations and watchkeepers. Managing the costs of switchtasking is increasingly important because technology offers tempting options as never before. Cell phones, text messaging, email, and other modes of instant communication all lend themselves to switchtasking, and they are all on boats.

Switchtasking has its place, as do the technologies that make it possible. You are switchtasking when you talk to another vessel on the radio while identifying its AIS signature and monitoring its relative motion. But in this instance all the activities serve your immediate responsibilities; none of them represents a pursuit of lower priorities or personal amusement. The ability to switchtask is integral to the concept of the *team-of-one* that typifies the one-person wheelhouse. At the same time, the notion that there are limits to what one person can handle points to the value of having a bridge team that can expand or contract with the demands of the situation.

Switchtasking is not always acceptable even when it is job related. If it leads to an accident, making chart corrections or preparing for a safety audit when you should be focusing on navigation and traffic isn't any better than listening to music or shooting the breeze.

Our predilection for switchtasking raises important questions and has broad implications for managing the use of some technologies aboard. How much switchtasking watchkeepers are allowed to engage in sets a standard for acceptable professional conduct. A junior mate or a deckhand watching a seasoned veteran transiting a narrows amid traffic while cranking tunes and talking to a pal on the cell phone might get the idea that this is how it's done. There are several problems with this. First, some people are better at switchtasking than others, but most people fancy themselves to be pretty good at it. Second, no one is uniformly good at it at all times. When you stir in a dollop of complacency, stress, fatigue, or some other ingredient that degrades situational awareness, even a good switchtasker is more exposed than he or she knows. Third, switchtasking can become such a habit that people cease to pay much attention to anything or anyone for sustained periods. A captain, mate, or deckhand who only half-listens to what is being said due to incessant switchtasking is going to miss something important eventually. Some switchtasking may be necessary, but too much is just too much. Determining just how much and which kinds are acceptable, and when, can profoundly influence the shipboard culture and the attitude toward watchkeeping.

Fixation

Fixation, another form of distraction, can be viewed as the opposite of switchtasking. Also known as *coning of attention*, fixation occurs when a person gets sucked into one task or interaction to the exclusion of all others, and in doing so loses overall situational awareness. The captain of the *Maria Asumpta* (see Chapter 3) most likely suffered from fixation. It started with his insistence on making his approach to their destination under sail. Rather than abandoning the attempt when the ship started losing ground to leeward, he reset the goalposts closer and closer to shore in the belief that it could still be done. His fixation continued even after the engines failed when, under great stress, he continued to try and sail clear of the cliffs. The vessel was clearly in extremis, but he gave no orders to ready an anchor, distribute life jackets, or transmit a Mayday call, and he attempted no alternate maneuvers. Photographs show the vessel passing through waters foaming with backwash from the cliffs, yet even at this point the captain still thought he was going to make it. Unchecked individuals are more susceptible to fixation than teams or individuals who are guided by plans and procedures. A robust team is more likely to temper the single-mindedness of any individual, while also bringing more ideas to bear. Since solo watchkeepers don't have the benefit of a team, fixation may be a particularly insidious threat.

Personal Distractions

Personal distractions are activities conducted on the job that have no connection to it. Some personal distractions are undertaken by choice: playing music, computer games, texting, and talking or reading about something unrelated to the job. Others, such as daydreaming, are unconscious acts. There is no good time for any of these things on watch, yet they have all happened. Some organizations make rules to discourage personal distractions on watch. After two serious collisions involving its own boats in 2009, the U.S. Coast Guard took steps to prohibit the use of cell phones and other electronic devices by its crews while operating a vessel. Others rely upon common sense, which, unfortunately, isn't always common in this area.

Even in the absence of the more blatant examples above, people's personal lives have enormous potential to distract them from professional duties. This is because most people's lives are significantly defined by relationships, interactions, and associations: family, friends, colleagues, and finances all impinge on our professional lives to some extent. Try as you might to leave your personal life ashore, these attachments and obligations do not stop at the gangway. When there is stress in the private life, it is even more likely to follow a person aboard. Due to the wonders of modern communications, private lives intrude on shipboard life as never before. Most would say that the ability to stay in touch is good for morale, but it also creates a new potential for distraction.

Professional Distractions

On any vessel many distractions are related to the ship's business. Professional distractions can be especially difficult to recognize and resist because, at some level, each one is worthy of attention. Mariners may attempt to deal with them with the best of intentions but get into trouble because the timing isn't right: a crew member comes along to discuss a maintenance issue just as the radar picture is coalescing in your mind, or you are about to make a Sécurité call, or a course change. Usually you brush it off and stay on task, but sometimes you don't. Even when you do brush it off you suffer a loss of focus that must be regained. As with switchtasking, there is a slight transaction cost for recognizing a nonurgent interruption for what it is. Good crew who are situationally aware seem to know when not to interrupt. Other people have to learn.

Professional distractions can also include fussing over a piece of nonessential equipment, planning future tasks, mentoring a newcomer, or tending to administrative duties (paperwork) all while trying to keep the vessel out of harm's way. A classic example of a professional distraction is a cell-phone call from the home office that comes in the middle of a docking maneuver. With both hands full of radios and throttles, you try to concentrate with the phone ringing annoyingly in the background. When all is secure, you call the office back to find out what they wanted and they reply, "We just wondered if you were there yet."

Some professional distractions seem to be structural—they are built right into the operation. The captain of the *Andrew J. Barberi* (see Chapter 1) was supposed to have been on the bridge with the assistant captain at the time of docking. Instead, he was elsewhere doing paperwork. Paperwork had become a structural distraction that led him to abandon the standard operating procedures on a regular basis. The first officer aboard the *Herald of Free Enterprise* faced a structural distraction because the schedule essentially required him to be in two places at once. He "solved" it by leaving the loading deck before the bow doors were closed. He was distracted by his next duty before he'd finished his previous one. Both tasks were relevant, but his priorities became reversed.

Professional distractions are like a Trojan horse. Because they fall within the scope of our duties, we accept them without suspicion. Structural distractions require structural solutions, which often requires an outsider to analyze the operation if the people inside the system lack the objectivity to recognize structural distractions. All distractions pose a threat, but the professional ones may be the most common and the best disguised.

Defenses against Distraction

If we always knew when we were succumbing to distraction, there would be no case studies on the subject. As it is, because people are imperfect observers of themselves, many of the human factors that degrade situational awareness are invisible to us in the moment. So it is with distraction. Distractions are a moving target since our vulnerability to them varies among individuals and with circumstances. In general, however, defending against distraction involves three processes: *Forehandedness*, preparing fully in all ways appropriate to the situation so as to minimize the unexpected; *Mindfulness*, continually taking stock of what is out there, where you are in it, and what *could* happen as a result; and *Prioritizing* in accordance with what you already know and with new information coming in. Here are some tools that can help.

Planning and procedures: Sound operating procedures that promote safe practices can help keep priorities straight, leaving less room for distractions to flourish. A passage plan, which is a kind of procedure, can reduce the need to divide attention between monitoring and assembling your plan on the fly, while also keeping priorities straight.

Bridge teams: As the *Queen of the North* incident demonstrates, a bridge team, something that is supposed to raise situational awareness, is capable of creating its own distractions. Flagrant violations of good watchkeeping aside, a well-tuned bridge team should be a positive force for resisting the effects of fixation, interruptions, and both personal and professional distractions.

Alarms and Indicators: Audible and visual alarms are important tools for prioritizing attention, but they can be as good as bad, and they often highlight the futility of attempting to regulate human behavior with things that beep and light up. Alarms primarily defend against neglect. They are intended to attract attention from one thing and redirect it to another presumably more important thing. Alarms are distracting in their own right, though presumably with a higher purpose. Anyone who has sailed with alarms

TIME-USE MATRIX

	URGENT	NON-URGENT

IMPORTANT	**I. Mission Central**
	• Keeping the schedule
	• Making critical operating windows: (weather, tide, current, daylight, ice)
	• Carrying out the mission with regard to cargo, equipment, passengers
	• Urgent communications
	• Correcting deficiencies
	• Obtaining critical parts, supplies
	• Dealing with crisis:
	  - Fire, MOB, injury
	  - Critical equipment failure
	  - Hazardous traffic or navigational situations
		II. Steady As She Goes
		• Planning
		• Drills and training
		• Preventive maintenance
		• In-house inspections and checks
		• Routine communications
		• Record keeping
		• Rest

NOT IMPORTANT	**III. Distraction Central** • Non-essential interruptions from office, crew, shoreside officials, and family via email, phone, etc. • Trivial problems presented as urgent • Unnecessary paperwork or tasks that are important to someone but not central to your mission • Low-priority communications • Other people's problems	**IV. Polishing the Binnacle** • Useless busywork • Non-essential maintenance • Recreation • Time wasters *These things are fine if nothing else needs to be done*

The highest priority of a watchkeeper is to stay focused on a safe passage. Being alert to what is truly urgent and important is integral to deflecting distraction and keeping priorities straight.
(Adapted from Franklin Covey)

knows that they have the potential to distract in ways that are not always warmly appreciated. This is especially true when an alarm malfunctions on a recurring basis. These are the times when it is tempting to disable or bypass them, as was done aboard the *Queen of the North*. For all their imperfections, the judicious use of alarms can help jolt a distracted or inattentive watchkeeper back to the task at hand.

Time management: Effective time management is part of prioritizing and keeping distractions at bay. The time-use matrix above shows the relationship between importance and urgency. This concept of time management is not a defense against short-term interruptions or poor judgment like that shown by the *Queen of the North*'s fourth mate, but it is interesting to consider that the more time a person spends in Quadrant II, the less time is spent in crisis (see Quadrant I). Time management is a worthy goal aboard any vessel.

Stress, fatigue, complacency, and transition: We explore these topics in the following chapters, but our vulnerability to distractions is deeply entwined with other factors that impact situational awareness. In particular, we should note that a number of recent transitions were in play aboard the *Queen of the North* at the time of the accident, any of which could contribute to below-average situational awareness: The crew had recently rejoined the vessel after a four-month refit; the personal relationship between the fourth mate and the deckhand had recently ended; the fourth mate had just relieved the second mate at the conn; the vessel was transitioning from confined waters to more open water, potentially inducing complacency; a course change was required; and the steering system was new and unfamiliar.

*****

In many instances, what makes distraction dangerous is not the distraction itself but rather its timing. Bad timing is difficult to defend against because it is frequently beyond our control. Even so, keeping priorities straight still comes down to individual choices and actions. You can manage distraction by being attuned to the many forms it takes, recognizing it for what it is, and acting accordingly, sometimes firmly.

Chapter Seven

Stress

STRESS AND DISTRACTION CAN FORM a vicious cycle. The more stress we are under, the more vulnerable we are to distractions. The more distractions around us, the more stressful a situation may become. Stress can make it difficult to focus, thereby becoming its own distraction. The emphasis here is on work-related stress, but obviously stress has many external sources that can find their way into the workplace. And even if *your* life is stress-free, that may not be the case for all of the people you work with, or for the person in the wheelhouse of the oncoming vessel. Other people's stress can become your stress, and we need to bear that in mind when thinking about human factors and situational awareness, not to mention leadership.

Stress is a normal part of life, and it is not always negative. Adventure, new experiences, competition, and challenges can generate stress that is stimulating and exhilarating. Negotiating a challenging stretch of water can be stressful, but if your skill is equal to the task, it can also be rewarding. Stress of the right kind helps you stay alert and engaged in your surroundings, which is important for a watchkeeper. Without any stress at all you might become complacent and apathetic. Lots of times a demanding watch is a good watch because you feel challenged, busy, and stimulated. But when stress levels get too high, particularly over a long time, we can become overloaded, and this can be destructive. Overload can lead to disorientation, either literal or figurative, and misplaced priorities. When this happens, stress clouds situational awareness and becomes a barrier to good decision making. People are not all the same, and one of the ways this shows up is in their relationship with stress and their ability to handle it. In the context of BRM, we need to understand stress because it can degrade situational awareness and separate us from our better selves.

The Towboat *Mauvilla*, 1993

Nothing beats a nice day on the water. With good visibility you can relax a little, enjoy your surroundings, and still have ample situational awareness. You can confirm your position and size up traffic, if any, at a glance. If only every day could be that way. Restricted visibility, on the other hand, creates stress, or at least it should. Even with radar and sophisticated navigational systems, operating in fog is stressful for two reasons, both of which are unsettling. First, it forces you to rely more heavily on instruments, which we know comes with the risk of overreliance. Second, you are still left with an incomplete picture of what is going on. In poor visibility you have to work much harder to sustain adequate situational awareness, and it may still fall short of what your eyes could tell you in better weather. This sense of uncertainty can produce anxiety, as well it should, in all but the most experienced, or the most complacent, mariner.

In the early hours of September 22, 1993, the towboat *Mauvilla* was northbound on the Mobile River in southern Alabama, pushing six barges. The crew consisted of a captain, a mate (known as a pilot on riverboats), and two deckhands. The pilot had the watch and was navigating by sight and by radar. He made no adjustments to the radar upon taking the watch, and he did not note the range scale it was on. As allowed by regulations, the vessel had no compass. Instead, it was fitted with a swing meter that could indicate the vessel's rate of turn but did not provide heading information. As regulations also allowed, there were no charts aboard; therefore landmarks could not be confirmed, nor was it possible to compare a radar image to anything but memory.

At the south end of Twelve Mile Island (mile 7 of the river) a slight haze developed. By the time the *Mauvilla* passed Catfish Bayou, the fog had thickened. Approaching the north end of Twelve Mile Island the pilot radioed the *Thomas B. McCabe*, a northbound towboat up ahead, and asked about the visibility. The operator of the *McCabe* said that visibility was down to zero. The *Mauvilla*'s pilot responded that he intended to look for a tree to tie off to, a common practice under the circumstances. As he arrived at the north end of Twelve Mile Island the fog "shut in tight."

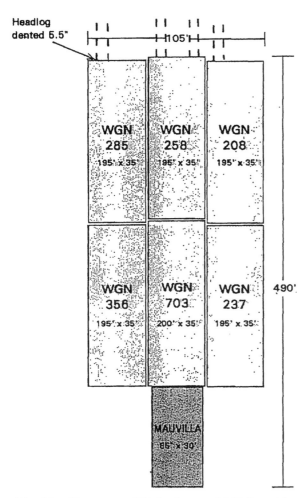

The *Mauvilla* tow was 490 feet long and 105 feet wide. On this trip it was pushing six barges in a two-row, three-column configuration. (National Transportation Safety Board)

The pilot ordered his deckhand to go to the head of the tow to snag a tree to tie off to. While maneuvering toward the bank, the pilot used his searchlight to try and spot a tree, but after several unsuccessful attempts he became concerned for the deckhand's safety and called him back. The pilot lost track of time during this interval and was uncertain how far he had traveled while looking for a tie-up spot. At one point he observed the swing meter indicating a turn to port, but he was too preoccupied to notice just how much. Having failed to find a place to tie up he decided to continue upriver to a more familiar tie-up.

Before long the pilot noticed a long, straight radar target that appeared to extend right across the river ahead of him. Since there were no bridges in this part of the Mobile River, he concluded that it must be another tow that had tied off to the bank and then swung perpendicular to the river. Thinking he would raft up to that tow, the pilot again contacted the *Thomas B. McCabe* to find out what the other operator knew about it, but the *McCabe*'s operator had not seen it. Approaching at what he believed to be "1 or 2 knots," the pilot twice radioed the stationary tow while steering, checking the radar, and working the searchlight, but got no response. As he nosed up to the target, the pilot felt a bump, as if he had grounded. The captain, who was asleep, felt it too and came to the wheelhouse. The duty deckhand, who was now sitting in the galley, felt it, and noticed that the clock read 0245. Two barges came loose, and the crew set about retrieving them.

At 0233, half an hour behind schedule, the Sunset Limited Amtrak train departed Mobile, Alabama, with 202 passengers aboard. The train was going 72 miles per hour as it approached the bridge that crossed the Big Bayou Canot waterway, about half a mile off the

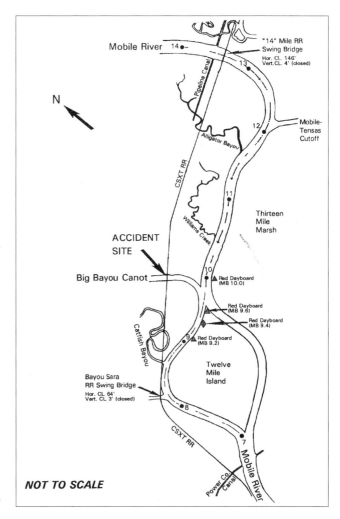

A depiction of the *Mauvilla*'s collision with the railroad bridge. (National Transportation Safety Board)

Mobile River, just north of Twelve Mile Island. Eight minutes after the *Mauvilla*'s tow struck that bridge, the train hit a displaced girder, plunged into the Big Bayou Canot, and burst into flames. The glow of the flames through the fog caught the attention of the *Mauvilla*'s crew, and the captain contacted the Coast Guard: "There's a hell of a fire up here in the middle of the river, and there ain't supposed to be no fire up here…I don't know exactly where we're at. It's so foggy I can't tell…there's something bad wrong up here." Forty-two passengers and five crew members were killed, and 103 passengers were injured. The towboat crew was not injured.

Too Much Stress

To unwittingly make a 90-degree turn off the Mobile River and into the Big Bayou Canot required a total breakdown of situational awareness on the part of the pilot. Not having a compass or charts certainly aggravated the pilot's disorientation in the fog. But it is also evident that he was too preoccupied to manage the resources that he did have (the radar; a swing meter; a deckhand, who was below instead of looking out; and the captain). The loss of situational awareness started with the fog, always a challenge, and worsened as the pilot's attention became increasingly divided between maneuvering, looking for a tie-up spot, directing his deckhand, and making radio calls in almost zero visibility, alone. As the pilot became overextended, he ceased paying attention to where he was going and his priorities became muddled. Finding a tie-up was important, but not more important than knowing where he was. Regardless of whether the pilot was consciously stressed or whether his grasp of radar and navigation was all it should have been, this trip was different because there were more things to do than he could do well. Investigators described him as "task-saturated." He became overloaded, and he lost the thread.

This tragedy was a defining moment for the inland towing industry. Sometimes that's what it takes. The licensing and examination structures were overhauled. Formal radar training, which the pilot lacked, and which had been required for bluewater deck officers for years, became mandatory for inland operators. Trained personnel are more likely to use equipment effectively than untrained ones, and well-trained personnel may experience less stress while doing so on account of their training.

The reforms that were implemented as a result of this incident were important, but to explain the accident as just the result of outdated practices obscures the human element, which is still very much with us. In the years since, countless mariners have found themselves overtasked, disoriented, and under stress with far better equipment than the *Mauvilla* had. The *Mauvilla*'s pilot was reasonably experienced: he had gone through an apprenticeship and had operated successfully on those waters at night before. He had not gotten this far on luck, but as long as people get into situations where there is more happening than they can handle, stress will shut down situational awareness.

Stress and Stressors

Stress is a natural response of the body to demands placed upon it, and it can have both biological and psychological elements. Events and circumstances that create stress are known as *stressors*. Dr. Hans Selye, a Canadian endocrinologist who did groundbreaking work on stress, developed the general adaptation syndrome (GAS) stress model, which has three stages.

The first stage is alarm. When your mind perceives a threat, or some other stress-inducing stimulation, adrenaline and other powerful chemicals are released into your body. These chemicals supercharge the system and prepare the body to take action. The alarm stage is rooted in a primitive "fight or flight" response to perceived danger that causes blood to flow from the brain to the muscles. A sense of anxiety or even fear may accompany this. Imagine being in a crowded anchorage when a sudden squall causes the anchor to drag. You leap into action and veer more chain. The anchor bites, and the situation comes back under control. The sense of alarm subsides, but as the chemicals released into your body dissipate, the experience may leave you feeling spent.

The second stage is resistance. Your body cannot sustain the high state of alert brought on by the alarm stage indefinitely. If the source of stress continues, the body attempts to cope. To continue with our example, perhaps veering more chain didn't work, so you start the engine to take strain off the cable. The situation stabilizes but you must remain at the controls, continuously making small adjustments of rpm and rudder while monitoring your position. There is a vessel dead astern, and other nearby vessels are dragging too. The

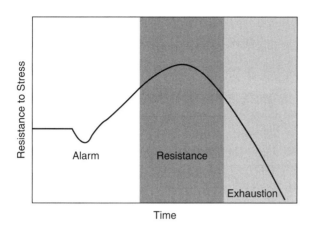

A graph of Selye's general adaptation syndrome.

Chronic Stress

Chronic stress develops over prolonged periods. It may be manageable, or it can build to levels that significantly impact performance and behavior. When a person is chronically stressed, the body has no chance to recover, and the individual becomes increasingly run-down. Sources of chronic stress aboard ship may include meeting the schedule, coping with regulations, keeping up with maintenance or administrative responsibilities, and dealing with people or the organizational culture, all the time knowing that if certain things go wrong, the consequences can be severe. We do not know what role chronic stress may have played in the *Mauvilla* case, but it is a fact of life for

situation is highly demanding and out of the ordinary. The source of stress does not go away. Externally, you are making a full effort to understand everything that is happening and make appropriate adjustments. Internally, your body is fully engaged in resisting the assault of stress on the system. But the resources for resistance are limited.

The third stage is exhaustion. If your body remains under stress, at a certain point its resources will be depleted and it will cease to function at a normal level. In our example, if you stay at the controls, there will be a mental and physical toll. As exhaustion sets in, reflexes, judgment, decision making, and overall situational awareness will suffer. You will start making mistakes.

Acute Stress

Acute stress is a response to an immediate perceived threat. It may be physical, emotional, or psychological; it may be real or imagined. It is associated with short-term situations, although it can happen on a recurring basis, such as negotiating a particularly harrowing stretch of water or dealing with a difficult individual. Acute stress can be brought on by emergencies, traumatic events, or a combination of factors that abruptly become overwhelming. The pilot on the *Mauvilla* may well have experienced some level of acute stress: everything was fine and normal, and then suddenly things piled up and situational awareness slipped away. The captain of the *Maria Asumpta* surely experienced acute stress at the moment his engines failed. In that case, last-ditch options existed, but they involved radical action, and a level of mental agility that he couldn't summon in the moment. Stress does this.

Signs of Stress

Individuals cope with stress in various ways. Some people appear easygoing, and actually are. Others appear easygoing but in fact are under stress, while others still are under stress and show it. Acute stress and chronic stress share many of the same indicators. Here are some warning signs:

- distraction
- disorientation
- disorganization
- mistakes, memory lapses, and oversights
- indecision
- slow responses
- poor judgment
- irritability and tension
- fatigue

In addition to the above, chronic stress may reveal itself in other ways:

- depression or general negative outlook
- inability to concentrate
- a feeling of being under tremendous pressure
- sleep disorders, including insomnia
- physical health issues, including diabetes and digestive and cardiovascular problems
- persistent conflict with other people: friends, family, shipmates
- substance abuse

The point of listing the signs of stress is to give a basis for observing and identifying its effects in everyday life. Stress is serious. Too much of it can lead to permanent long-term health issues and accidents.

many people, including mariners. Rather than ignoring it, or hoping it will go away, let's place the matter where it belongs: squarely in the middle of other factors that are known to degrade situational awareness. This way there is at least a chance of recognizing its effects on ourselves and our shipmates, so that some effort can be made to manage it.

Managing Stress on Watch

Stress is partly a function of *perceived* risk; therefore it often arises around activities that seem risky or threatening. But what one person considers risky or threatening, another may not; this just means that specific causes of stress vary with the individual. The causes of chronic stress are embedded in everyday life, often right under our noses, further complicating our efforts to manage it. The following section introduces general conditions that contribute to stress and offers some countermeasures, many of which fall well within the scope of good bridge resource management.

Inadequate bridge resources: Stress can develop from trying to meet goals or expectations with inadequate resources. What constitutes adequate resources depends on what you are attempting to do and the prevailing conditions. For the pilot on the *Mauvilla*, negotiating the Mobile River was beyond his abilities that night. Failing to consider alternatives, people sometimes attempt to bridge the gap between goals and resources on their own. While everybody respects a hard worker, when such efforts induce stress and a loss of situational awareness, those efforts are not a good idea. On the *Mauvilla*, the captain and the on-duty deckhand were the most immediate resources that went untapped. Whether the standard operating procedures were not clear enough or the operational culture discouraged an early call to the captain is irrelevant now. The pilot was overloaded and alone. The story of the *Mauvilla* makes a strong case for bridge teams in certain situations, even on vessels that don't normally use them. Had there been a team, this tragedy would almost certainly have been averted.

Time and speed: Time is a resource, and time available influences the speed at which a vessel travels. The faster a vessel travels, the faster its operator must process information and the less reaction time he or she has. Depending on the waterway, the conditions, and the number and complexity of other tasks that must be performed at the same time, the speed of advance can raise stress levels. Stress, therefore, can be a function of speed. In a given stretch of water, a watchkeeper may be perfectly effective at 8 knots and a nervous wreck at 12. Minimizing the effects of excessive stress—confusion, indecision, disorientation, poor judgment, etc.—is simple: SLOW DOWN. Or stop. Obviously, there are practical constraints to this edict, but it is one of life's great truths that, for a given situation, the faster you work, the more mistakes you make. As one captain put it, "The faster I go, the dumber I get." The pilot on the *Mauvilla* was actually trying to stop when the towboat struck the bridge, which only shows how fast things can go wrong once they start to go wrong.

Time and place: In the military arena, being able to choose the field of battle (when and where to fight) is a great advantage. For sailors, that advantage is known as the *weather gauge*. Controlling when and/or where higher risk activities occur is critical to a watchkeeper. A routine port-to-port meeting with another vessel can become a career-ending nightmare if it occurs at a bend in the channel just as a third vessel decides to overtake. If, through a speed adjustment, that same meeting happens on a straightaway and doesn't involve a third vessel, stress and risk can be minimized. The same is true for transiting difficult stretches of water in daylight versus darkness, at slack water versus maximum current, and so forth. You cannot always choose the field of battle, but where you can, you should. Passage planning is a process that can help us choose at least some of our battlefields.

Lack of continuity: Unfamiliar vessels, waters, operations, and shipmates can raise stress levels for anyone. Without the benefit of practice and familiarity, everything is harder. In time, continuity makes those same activities far less stressful. The *Herald of Free Enterprise* had the right number of crew aboard with the appropriate credentials, but the frequent turnover of officers interfered with the development of continuity and placed additional pressure on the regular crew. Transitional periods are unavoidable, and change can actually stimulate situational awareness. But to anticipate and prevent the effects of stress, mariners should pay special attention during times of discontinuity.

Lack of training and/or experience: For inexperienced crew, continuity does not yet exist, and many aspects of the job may be stressful. Fortunately, inexperience doesn't last forever, but when it is known to be a factor, extra thought should be given to how resources and responsibilities are distributed.

Good training, applied as intended, is a powerful bulwark against stress. People who have been taught the correct way of doing things will be more comfortable than those who haven't. Studies of Army Green Berets and Navy SEALs have shown that in threatening situations they initially experience about the same rush of the stress hormone cortisol as ordinary soldiers, but the levels drop much faster than they do among less well trained troops, leaving them less depleted and more able to focus on countering the threat. The *Mauvilla* pilot's lack of radar training meant that he was challenged both technically and in terms of managing his stress.

Fatigue: Stress can induce fatigue and fatigue can magnify stress. Where one can be anticipated, the other may also occur. We have all experienced the way in which fatigue saps our ability to handle fresh challenges. Under these circumstances even routine tasks may feel stressful. Strategies that minimize fatigue can have benefits for managing stress. We will look at fatigue in more detail in the next chapter.

Workload: Most of the issues above have workload as a common denominator. A balanced, manageable workload can help control stress; conversely, an excessive workload can increase it, both in the short term and the long term. Aboard the *Mauvilla*, overtasking led to diminished situational awareness on short notice. Determining appropriate workloads is not usually the province of individual mariners, but in general, the effect of excessive workload can be mitigated by the addition of resources of one type or another. There will always be times when mariners have to make do; the ability to improvise is a valued trait among mariners everywhere. But when making do becomes the norm, stress may become part of that norm, and serious consequences can result.

Personal and Professional Stress

This is a professional book, so I have focused the discussion of stress on themes that pertain to watchkeeping performance. As noted, however, mariners' private lives can be a source of stress. Worries pertaining to personal relationships, finances, and the mere fact of separation are well known sources of stress in the marine workplace. Sometimes stress appears self-inflicted through workaholic tendencies, martyr complexes, or personal ambition. Such issues are beyond the reach of this book, but it's commonly accepted that personal strategies for managing stress include regular exercise, a healthy diet, and healthy lifestyle choices. Illness of any kind stresses the body and its systems, so the more time a person can stay healthy, the better off he or she is with regard to stress.

Remember our discussion of switchtasking in Chapter 6? Many people feel like they don't have enough time to do everything they are expected to do, both on and off the job. Effective time management can help reduce some of that stress. Refer to the time-use matrix introduced in Chapter 6 to help prioritize the tasks you need to accomplish.

Personality clashes can be a source of stress anywhere, but they may become more intense when there is a pronounced hierarchy and no way to get away from the source of irritation—like on a ship. The causes of personality clashes are so varied that prescriptive remedies are impossible, but finding ways to de-escalate such tensions is usually in everyone's best interest, and is a mark of leadership.

Substance abuse, sometimes known as self-medication, is a well-traveled path in the quest to manage stress. Not only is it illegal on ships, but it makes matters worse over time.

In today's world, technological advancements often set the pace for individual performance. Technological upgrades can lead to safer, more productive operations while expanding the realm of what is humanly possible. But better technology can also lead to overreaching or to overestimating what people can reasonably accomplish. Better technology often means operating more aggressively by increasing the speed of advance under poor conditions, or decreasing turnaround times, or performing the same tasks with fewer people—or with the same number people but then giving them additional responsibilities. Or worse yet, doing everything faster with fewer people who now have more responsibilities. At some point the innovations that should have made the job safer and less stressful can be submerged by higher expectations. As a result, people may end up functioning closer to the limits of their natural abilities than ever before, despite—and because of—better tools than ever. When people function at or near the margins of their abilities on a regular basis, accidents will occur. Who will it be? Where will it happen? What will be the circumstances? This is the train of thought that, after all these years and all the changes, makes the *Mauvilla* incident an important story to tell and an important lesson to remember.

*****

Mariners are expected to handle what comes their way. When aboard, we are not paid to relax and pursue personal fulfillment. But by looking at time off and time on as two parts of a whole, it is possible to strike a reasonable balance in the work-play equation that keeps stress at manageable levels. If you cannot manage your own stress, then you cannot set an example that will help other people manage theirs.

Managing stress starts with self-awareness, in other words, situational awareness turned inwardly. Managing stress requires us to ask and answer the question, what is *my* situation? The "fight or flight" reflex that underlies much stress is fine for outrunning or battling predators, but in the execution of our normal duties it mostly interferes with cognitive control—the ability to think straight, maintain perspective, and react appropriately. Any technique that helps us maintain cognitive control can help us manage the negative aspects of stress. When demands begin to outpace our abilities, two countermeasures we may want to remember are *getting more resources* and *slowing down*.

Chapter Eight

Fatigue

According to a six-year study carried out by Cardiff University in Wales, one in four seafarers has reported falling asleep on watch. And almost half of all seafarers reported working 85 hours a week or more, and that their work hours had actually increased despite international regulations designed to ensure adequate rest. Throughout the transportation industry, fatigue is a widely acknowledged concern. In the marine sector alone, trade organizations, safety boards, and regulatory agencies (extending all the way to the IMO) have grappled with the issue. However, grappling with an issue and actually making it better are two different things. Suffice it to say, the extent of the dialogue over fatigue is an indicator of the magnitude of the issue, and the far-reaching policy implications it represents.

In the more immediate sphere of the workplace, anyone in charge of a watch or a deck team should be concerned about the potential for fatigue-related accidents. People in leadership positions need to anticipate, mitigate, and manage fatigue in others as well as in themselves.

Traditionally, fatigue has been thought to be the result of a lack of rest or doing a task for too long. Yet, understanding fatigue requires looking beyond a tally of hours worked or not worked in a day, week, or month. The duration of rest periods, the quality of rest, natural body rhythms, and the effects of light, energy intake and expenditure, and stress are all necessary to consider in any fatigue management approach. These issues are relevant to any industry defined by shift work, but shipboard fatigue involves still other factors including the following:

- Marine operations do not take place in a controlled environment. The effects of heavy weather, noise, vibration, temperature extremes, and constant motion are often inescapable whether a person is on watch or off.
- Manning levels are generally kept to the minimum required, leaving little crew to spare when someone is temporarily incapacitated. The mobile nature of a vessel makes it difficult to suspend operations or augment crew when this happens. In other words, you can't just pull over for a nap or turn your duties over to someone else if you start to nod off. The same goes if performance is impaired by illness or some other short-term condition.
- The demands of the watch cycle, cargo operations, safety drills, maintenance, and passenger and administrative duties reinforce the reality that on a ship, the show goes on for better and worse.
- Work does not have to be physical to be draining, and watchkeeping can be intense work. Navigation requires avoiding hazards in every direction, including above and below. Environmental factors (wind, tide, current, and visibility) and other variables require constant analysis, judgment, and decision making. Trips may be similar but they are never the exactly same. While this can alleviate problems associated with repetition, it can also contribute to fatigue.
- Shipboard life often involves prolonged separation from home and personal responsibilities, which can cause chronic stress and aggravate fatigue.

Fatigue and Casualties

Unlike drugs or alcohol, there is no universally accepted pass-fail test for fatigue. Measuring fatigue is difficult, and pinpointing its precise role in any accident with confidence can be tricky. The lack of "proof" can lead some to dismiss the role of fatigue when it may well have been a factor. Occasionally, fatigue is plainly causative. For instance, on February 15, 1985, the captain of the passenger ferry *A. Regina* grounded on Mona Island, between Puerto Rico and the Dominican Republic. The ship, valued at $5 million, was a total loss. Mona Island presented a large and unmistakable radar target and it was the only land for many miles around. But the captain hadn't had a day off in a year, and a new schedule had just been put into effect that further reduced downtime.

These conditions, combined with insomnia, meant he had not slept for 42 hours before the accident.

In another example, on June 29, 2003, the mate of the small coastal freighter *Jambo* ran the ship aground on a rocky outcrop near Ullapool, Scotland, and sank it. He missed a course change at 0415 and awoke an hour later when the ship struck the rocks with him standing at the controls. The watch rotation was 6 hours on and 6 hours off, but frequent port calls and the need to supervise cargo operations in port made it "6 and 6" only in name. Neither the mate (nor the captain or their employer) gave sufficient weight to the risks of falling asleep on watch. The mate made matters worse by allowing his AB to leave the bridge for extended periods to smoke. This in itself demonstrated the effect fatigue may have had on his judgment, and, as a result, no one was there to see him fall asleep. These are obvious cases of fatigue, but most cases are less clear-cut.

For instance, when the ferries *North Star* and *Cape Henlopen* collided in Long Island Sound on July 9, 1987, dense fog was the primary reason both captains lost situational awareness. However, the captain of the *Cape Henlopen* had been working 16 to 17 hours a day for the previous four days, only getting 5 to 6 hours of sleep a night. Not unlike an intoxicated person claiming he/she is fine to drive, the captain claimed he didn't feel tired at the time. That doesn't mean he wasn't fatigued. The captain's comment says a lot about the hazards of self-diagnosis and how fatigue can interfere with judgment. It is impossible to say how much of the captain's lost situational awareness was due to fatigue, but common sense tells us that his work pattern was detrimental to getting adequate rest. The comparison between the effects of fatigue and alcohol consumption is not casual. A study published in the July–August 1997 issue of *Nature* magazine showed that 17 hours of sustained wakefulness produced a level of cognitive psychomotor impairment equivalent to a blood alcohol level of 0.05. IMO standards prohibit a blood alcohol level greater than 0.04 on duty.

The assistant captain of the Staten Island ferry *Andrew J. Barberi*, who blacked out at the controls, reported afterward that he was "exhausted at the time, but no more exhausted than usual." He had a regular shift but worked overtime as well. He was taking five medications, including a narcotic for pain and another drug for insomnia, each with significant potential side effects. At home, an eight-month-old baby had been up during the night prior to the accident. The assistant captain may have been suffering from chronic low-level exhaustion. If he was, was it a factor, and how big a factor? We will likely never know, but we should bear this man's situation in mind when we look at the complexities of our own lives and try to evaluate the potential for fatigue (and stress and distraction) to affect shipboard performance.

The role of fatigue in accidents often rests on indirect or circumstantial evidence, which is inherently hard to pin down. A 1996 U.S. Coast Guard study of 279 incidents showed that fatigue was a factor in 16% of critical vessel casualties and 33% of personal injuries. This is not insignificant, but many believe that the role of fatigue is far greater than these figures suggest.

Let's go back and reexamine the *True North* case (see Chapter 5) to see how fatigue may have worked behind the scenes to add to the captain's overreliance on the electronic navigational system.

The *True North* and Fatigue

The *True North* grounded in a remote location on the northwest coast of Australia when the electronic chart system gave the captain a faulty picture of the vessel's actual position. Radar, and possibly other sources, would have revealed the discrepancy, but the captain did not aggressively cross-reference the ECS picture. Investigators decided to analyze his work patterns to determine the role of fatigue in his performance. What they found was interesting, and it may say much about how fatigue creeps into everyday life unnoticed.

Like many smaller commercial vessels, the *True North* carried just two licensed watchkeepers, the captain and a mate. Naturally, the captain's duties extended beyond his hours of watchkeeping and included socializing with passengers. By 2300, the time of the grounding, the captain had been active since 0600, with the exception of a break just prior to taking the watch at 2200. This is a long workday by any measure but not unheard of. The captain had been following a similar routine since taking command eight days earlier yet he testified that he had not felt tired when he took the watch. For a number of reasons, this statement is not surprising.

Though there is no universally accepted definition of fatigue, one study carried out by the Australian government in 2000 distinguished between *tiredness*, the ability to initiate sleep, and *fatigue*, which is based on the ability to maintain adequate job alertness. By these definitions, it would be possible to be fatigued without necessarily feeling ready for sleep. The captain's remark also resurrects the question of objectivity in gauging your own condition. In fact, it is not uncommon to find a discrepancy between how people *say* they feel and the level of alertness they show when tested. A 1986 study of airline crews found that some individuals who had evaluated themselves as being at maximum alertness were actually falling asleep within 6 minutes.

As part of the *True North* investigation, the work-rest patterns of the captain were analyzed using fatigue audit software from the Centre for Sleep Research at the University of South Australia, which generated a score based on several aspects of his recent work history. The scoring system was as follows:

Less than 80	Work-related fatigue unlikely
80 to 100	Some people will show signs of fatigue impairment on some tasks
Greater than 100	All people are likely to be fatigue impaired by any task

Research by the CSR suggests that a score of 80 to 100 is high, and 100 to 120 is very high, and an additional study found that a score between 80 and 100 "is comparable to . . . a blood alcohol content of 0.05% or greater."

The software employed two different scenarios to derive a score for *True North*'s captain. The first scenario was based on hours of actual work, and the second was based on all waking hours. In the first situation, the captain received a score of 97, the high end of "high." The second scenario produced a score of 117, the high end of "very high." Even assuming that the reality was somewhere in between, we are still left with a strong case that fatigue played an important role in diminishing his situational awareness and in amplifying his tendency to overrely on the easiest monitoring tool available, the electronic chart. This may well be an example of what mariners typically face: fatigue that is not debilitating, but which renders them nowhere near their best.

Signs of Fatigue

Stress and fatigue are different, but they can cause similar symptoms: irritability, indecision, an inability to focus, a tendency to fixate, poor judgment, and so on. In addition to these signs of diminished situational awareness, signs of fatigue can include the following:

Memory lapses and errors of attention: Fatigued individuals may forget steps in a sequence or overlook key details. They may also neglect to consult the checklists and operating procedures that could assist them.

Cutting corners: Fatigued individuals may select strategies or make high-risk choices because they seem to involve less work. The doctrine of "good enough" expands into areas it should not.

Poor communication: High-quality communication requires effort. Fatigue can cause people to be less communicative or communicate less effectively. They may become withdrawn, which may make other team members uncertain and less effective.

Delayed reactions, both mental and physical: Fatigue can slow reaction times. For instance, when sizing up a traffic situation, fatigued individuals may have trouble reacting correctly in a timely fashion.

Reduced vigilance, lethargy, and even apathy: In severe cases, fatigued individuals may not react at all. The mate on the *Jambo* actually fell asleep at the controls.

Managing Fatigue

Torbjorn Akerstedt, professor of behavioral medicine at Stockholm University and leader in fatigue research, has described fatigue as "the tiredness and sleepiness that results from insufficient sleep, extended number of waking hours, and circadian rhythms." Managing fatigue requires a basic grasp of the natural requirements for rest and how the body tries to regulate energy on a daily basis. The term *circadian* comes from the Latin words *circa*, meaning "around," and *dies*, meaning "day." Thus, the *circadian rhythm* describes the natural ebbs and flows of alertness and tiredness, and the release or replenishment of energy in a 24-hour day. Because the circadian rhythm is regulated by cycles of daylight and darkness, it is always at work, regardless of your work schedule. Obviously, the circadian rhythm does not always mesh with the watch rotation. This can result in reduced alertness during certain hours of the day, even when watchkeepers *have* had adequate sleep.

The circadian rhythm produces two peaks and two troughs of alertness each day. In the typical pattern, energy levels rise through the morning after waking up and reach a peak at around 1000. Energy and alertness then drop off gradually, bottoming out somewhere in the late afternoon. In the early evening, alertness rises again and peaks for a second time before declining through the hours of darkness and reaching a nadir between 0300 and 0500. This nadir is when most people on a typical shoreside schedule experience their deepest sleep. One fatigue study has designated the period from 2100 to 0700 as the *red zone*, the most critical part being from 0300 to sunrise. That is when energy and alertness are at their lowest and most likely to impact people who are working, even if they are well rested. The circadian rhythm can, of course, vary among individuals, and it can also fluctuate with the seasons, but in general, the red zone should be viewed as a period of elevated risk for fatigue-related incidents.

The circadian clock, or body clock, is a powerful regulator of human performance. Left to its own devices, it helps us use and replenish energy efficiently, but there is little about shipboard life that supports these natural rhythms. Changes to established work-rest schedules and time zone changes disrupt the circadian rhythm, causing the body clock to desynchronize. The body can take several days to resynchronize, and even then it will resynchronize only if the new pattern is maintained. Inconsistencies in work-rest patterns (as opposed to changing from one routine to a different routine) can keep the body clock off balance indefinitely. If the watch cycle is at odds with the circadian rhythm, it may put a strain on the body no matter how regular the routine. These are all commonplace circumstances that can interfere with the body's ability to regulate energy effectively. People struggling with a desynchronized body clock may appear able to carry out their assignments, but physical coordination and vigilance will suffer. The mate on the *Jambo* had a regular routine on paper, but it was routinely disrupted by cargo operations and daywork in port.

The circadian rhythm can moderate or exaggerate the effects of sleep deprivation. Exhausted people can experience a misleading and temporary energy boost during a circadian peak that can convince them they are alert when, in fact, they may not be. Conversely, a circadian trough may amplify the negative effects of inadequate sleep, making people a real danger to themselves and others. The mate on the *Jambo* was already sleep deprived when he came on watch, but standing the midnight to 0600 watch did not help. He drifted off during the deepest part of the red zone.

Sleep Cycles and Sleep Fragmentation

During normal sleep, the brain cycles through periods of dream (REM—rapid eye movement) sleep and non-dreaming sleep (NREM). It takes around 90 minutes to complete the entire cycle. In the course of a full night's sleep, the cycle repeats several times. If a person is awakened mid-cycle and goes back to sleep, the brain starts the sleep cycle from the beginning, cutting short the benefits of the full cycle. If a person is awakened several times in a sleeping period, as can easily happen aboard, *sleep fragmentation* can seriously disrupt the rejuvenating benefit of dream sleep without appreciably reducing the total rack time. When interruptions occur, a person may awake feeling tired despite having had several hours sleep.

Sleep duration is critical to the body's ability to restore energy. *Adenosine triphosphate* (ATP) is a natural catalyst that takes energy from food and transforms it into usable energy. The production of ATP depends on a number of things, but chief among them is sleep. There is no substitute for ATP and no artificial source. Research shows that on average, people need about 8 hours a day of *continuous* sleep to produce enough ATP to meet the body's daily energy needs. Based on this premise, two 4-hour sleep periods do not equal one 8-hour period, a fact that presents a built-in fatigue issue for the "6 and 6" watch rotation. After tending to off-watch duties and personal requirements (eating, laundry, hygiene), it can be difficult to obtain even 5 hours of uninterrupted sleep. Though some lab research suggests an *anchor sleep* supplemented by *strategic napping* may be equivalent to an 8-hour consolidated sleep, this is not firmly established. In any event, if the body doesn't produce enough ATP, signs of fatigue will appear.

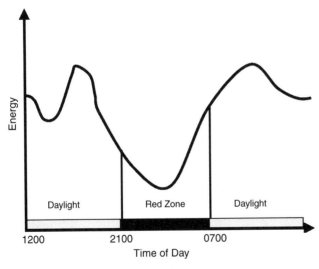

The red zone of performance refers to the hours from 2100 to 0700, when human performance (energy and alertness) are at their lowest. The afternoon trough is not as pronounced as the early morning one. (United States Coast Guard)

Sleep Debt

When a person chronically functions with less than 7 to 8 hours of sleep a day, the effect is cumulative and leads to *sleep debt*. If a person requires 8 hours sleep but only gets 5, he or she incurs a sleep debt of 3 hours. If this pattern continues for four days, as it did for the captain of the ferry *Cape Henlopen*, the sleep debt would be about 12 hours. Strategic napping is increasingly accepted as a way to alleviate sleep debt, but the timing and duration of naps is important in order to avoid working at cross-purposes with the circadian rhythm. Even when aligned with the 90-minute sleep cycle, naps are no substitute for consolidated rest. Just as sleep debt is cumulative, so is the impact it has on performance and alertness.

Recovery from sleep debt is not achieved on an hour-for-hour basis, but through an increase in deep sleep above the body's normal requirements. This recovery can be difficult or impossible to achieve within the confines of many watch rotations.

Sleep Disorders

Sleep disorders are much more common than many people realize, partly because people are often asleep when the disorders manifest themselves. When people are oblivious to having a sleep disorder, they may live with a degree of fatigue for years without realizing how differently they would feel if they were well rested. People are born with certain sleep disorders, while others develop over time. Obstructive Sleep Apnea (OSA), which interferes with breathing while asleep, has become increasingly recognized in recent years. Some sleep disorders may be aggravated by the watch cycle, others less so. This is not the place to catalogue specific sleep disorders in detail, but it is a serious enough matter that some organizations even test employees to determine if there is a hidden issue that could contribute to workplace fatigue. A great many sleep disorders can be managed and even cured, producing big quality-of-life improvements. Being overweight or in poor physical condition often aggravates sleep disorders, a point that further illustrates how a healthy lifestyle can play a part in a healthy career.

Beyond Sleep

The discussion of fatigue tends to focus on work-rest patterns, and rightly so, but research shows that fatigue management can involve lifestyle factors including diet, exercise, and stress management. Since much of fatigue management centers on replenishing energy, a balanced, healthy diet is important. The timing of nutritional intake also matters. The heaviest meal of the day should come after waking up from the longest sleep of the day; heavy meals before sleep should be avoided. Staying hydrated is one of the best, simplest steps a person can take to help the body work efficiently. As far as this goes, there is no substitute for water. All other beverages require the body to work harder for less benefit, and some are downright counterproductive.

Caffeine is effective in temporarily dispelling drowsiness, but like all such measures, you can overdo it. Too much caffeine can interfere with concentration while you're on watch and with getting to sleep when you're off watch. It can also heighten stress-related anxiety. Common caffeine sources such as coffee and tea are also diuretics and contribute to dehydration.

It is critical to thoroughly understand the intended effects and side effects of all medications being taken. Both prescription and over-the-counter medications, separately or in conjunction with one another, can interfere with alertness and sleep quality. This may have been a factor for the assistant captain on the *Barberi*.

Chronic stress can interfere with getting quality rest, thus highlighting the relationship between stress and fatigue. Exercise is a known stress reliever that can help manage chronic stress and thereby mitigate fatigue. Regular exercise, even in limited amounts, can help the body's efforts to produce and regulate energy. Adequate exercise can be hard to get on board, but some organizations have attempted to meet this need by installing exercise equipment.

Acute stress, such as might be experienced during an emergency, can leave individuals or an entire crew utterly depleted. In this situation, people may be at higher risk of making further mistakes that compound into an accident cascade. From a leadership standpoint, a crew that has undergone a stressful experience is a vulnerable one. If at all possible, an effort must be made to provide recovery time.

Fatigue Regulations

Regulations concerning work-rest minimums are crafted in distant locations, and management decisions regarding manning, workload, work rotations,

Fitness for Duty

Section A-VIII/1 of the STCW Code contains important requirements governing the fitness for duty of watchkeepers. (As per the Manila Amendments of 2010, these will become effective January 2012.)

1. Administrations shall take account of the danger posed by fatigue of seafarers, especially those whose duties involve the safe and secure operation of a ship.
2. All persons who are assigned duty as officer in charge of a watch or as a rating forming part of a watch and those whose duties involve designated safety, prevention of pollution, and security duties shall be provided with a rest period of not less than a minimum of 10 hours of rest in any 24-hour period, and 77 hours in any 7-day period.
3. The hours of rest may be divided into no more than two periods, one of which shall be at least 6 hours in length, and the intervals between consecutive periods of rest shall not exceed 14 hours.
4. The requirements for rest periods laid down in paragraphs 2 and 3 need not be maintained in the case of an emergency or in other overriding operational conditions. Musters, firefighting and lifeboat drills, and drills prescribed by national laws and regulations and by international instruments shall be conducted in a manner that minimizes the disturbance of rest periods and does not induce fatigue.
5. Administrations shall require that watch schedules be posted where they are easily accessible. The schedules shall be established in a standardized format in the working language or languages of the ship and in English.
6. When a seafarer is on call, such as when a machinery space is unattended, the seafarer shall have an adequate compensatory rest period if the normal period of rest is disturbed by call-outs to work.
7. Administrations shall require that records of daily hours of rest of seafarers be maintained in a standardized format, in the working language or languages of the ship and in English, to allow monitoring and verification of compliance with the provisions of this section. The seafarers shall receive a copy of the records pertaining to them, which shall be endorsed by the master or by a person authorized by the master and by the seafarers.
8. Nothing in this section shall be deemed to impair the right of the master of a ship to require a seafarer to perform any hours of work necessary for the immediate safety of the ship, persons on board or cargo, or for the purpose of giving assistance to other ships or persons in distress at sea. Accordingly, the master may suspend the schedule of hours of rest and require a seafarer to perform any hours of work necessary until the normal situation has been restored. As soon as practicable after the normal situation has been restored, the master shall ensure that any seafarers who have performed work in a scheduled rest period are provided with an adequate period of rest.
9. Parties may allow exceptions from the required hours of rest in paragraphs 2.2 and 3 above provided that the rest period is not less than 70 hours in any 7-day period. Exceptions from the weekly rest period provided for in paragraph 2.2 shall not be allowed for more than two consecutive weeks. The intervals between two periods of exceptions on board shall not be less than twice the duration of the exception. The hours of rest provided for in paragraph 2.1 may be divided into no more than three periods, one of which shall be at least 6 hours in length and neither of the other two periods shall be less than one hour in length. The intervals between consecutive periods of rest shall not exceed 14 hours. Exceptions shall not extend beyond two 24-hour periods in any 7-day period. Exceptions shall, as far as possible, take into account the guidance regarding prevention of fatigue in section B-VIII/1.
10. Each Administration shall establish, for the purpose of preventing alcohol abuse, a limit of not greater than 0.05% blood alcohol level (BAC) or 0.25 mg/l alcohol in the breath, or a quantity of alcohol leading to such alcohol concentration, for masters, officers, and other seafarers while performing designated safety, security, and marine environmental duties.

In addition to IMO guidance above, a number of initiatives worldwide have adopted more progressive approaches to managing fatigue. One of these is the Crew Endurance Management System (CEMS) developed by the U.S. Coast Guard in conjunction with industry. CEMS is a voluntary program that has been embraced and implemented to varying degrees. In addition to work-rest guidance, CEMS highlights the importance of the lifestyle choices discussed above. It provides direction for configuring vessels to be CEMS compliant by providing sleeping quarters, work conditions, and watch rotations that are believed to improve the quality of rest and minimize fatigue. CEMS implementation involves developing a vessel-specific plan, training crew members in CEMS practices, and then monitoring progress. For CEMS or any similar initiative to succeed, crew members and organizations must be given a realistic opportunity to change old habits and take advantage of an improved working climate.

and the allocation of responsibilities are decided well upstream from the typical mariner. Company culture and the degree of commitment to proactive policies in this area vary widely. The IMO "fitness for duty" standards apply to international operations (see the sidebar on page 93), and provide guidance worldwide, but the extent to which they are adhered to is unclear; many mariners simply smile when these standards are mentioned.

*****

The difficulty in getting adequate rest aboard a busy ship still leaves a number of things up to the individual. Getting rest when you can and for as long as you can is at least a partial choice that involves making good decisions about how to use off-duty time. Watching videos instead of catching up on sleep is not a good choice as far as this goes. Developing a routine and sticking to it helps avoid working at cross-purposes to the body's natural rhythms. While people may tire of hearing it, a healthy lifestyle is a significant defense against fatigue and brings many other benefits with it. Comprehensive fatigue management is a long-term proposition that involves changing ingrained habits and assumptions about how marine transportation should work.

Fatigue is a cost of doing business that is usually related to keeping the cost of doing business down. As with other human factors that depress situational awareness, whether fatigue is dangerous is a matter of degree: What's at stake? What are the risks? The day when all vessels operate with thoroughly rested crews at all times may never come, and this has much to do with forces and facts that are outside the control of anyone. But we can endeavor to pay closer attention to the issue and be proactive in managing it. When it is not possible to prevent fatigue, we can at least make informed operational decisions so that we don't expose ourselves or innocent people to danger.

Chapter Nine

Complacency

WE TEND TO ASSOCIATE COMPLACENCY with a variety of unflattering human characteristics—laziness, cockiness, sloppiness—and sometimes it is all those things. But ordinary complacency, the kind we are most likely to experience, is often a far more innocent phenomenon. Sometimes complacency is a cruel reward for actually becoming good at something. Sooner or later, everybody succumbs to it, though some are more susceptible than others according to circumstance and disposition.

Shipboard complacency arises when an attitude of contentment or satisfaction toward our responsibilities meets up with risk or danger. Sometimes complacency takes the form of a lack of vigilance and failing to notice important details or trends. Other times it takes the form of decisions that embrace risks that have not been assessed correctly. This occurred when the captain of the *Anne Holly* proceeded upriver with no assistance, and high water and darkness all working against him. In all cases, complacency reduces situational awareness, and it is a significant source of human error in the workplace. To understand complacency we must have a basic knowledge of how at-risk behavior shows up in the shipboard environment.

At-risk behavior has been characterized as being *intentional, unintentional,* or *habitual*. *Intentional at-risk behavior* occurs when people make a conscious decision to do something they know is contrary to established safe practices. They are not trying to cause an accident, but their choices make one more likely, and their choices are informed—they know better, and do it anyway. Some examples of intentional at-risk behavior are trying to make up time by operating at high speeds in foggy, congested waters, not conforming to the Rules at the Road, not following a passage plan, disobeying standing orders, or neglecting to repeat commands. In these situations, extraordinary ability is not required, only operating within established norms. Overconfidence and overfamiliarity are usually ingredients in this form of complacency.

Unintentional at-risk behavior is a fundamental lack of situational awareness caused by inexperience or lack of training. If people don't have a clue as to the hazards, they cannot ask the reasonable "what ifs"—they don't know better, and that's why they act in a risky manner. Examples of this include standing in the way of a line under tension, attempting an improbable maneuver, getting set onto a buoy—all out of ignorance of the forces involved. Being overconfident can play a role in this too, but it has a different source.

Habitual at-risk behavior is a product of repeatedly engaging in risky activity, intentionally or unintentionally. When people get away with mistakes, shortcuts, or a cavalier approach to danger, they cease to discern the risks: "I've done this a million times, and nothing has ever gone wrong." When things get to this point, methods and decisions may be based on a fortunate past rather than the actual risks. Often it is an utterly atypical, perhaps even bizarre, circumstance that abruptly exposes the truer risk. Examples that we've previously discussed of what some people may prefer to call "bad luck" include the following:

- The assistant bosun sleeping through the call to duty on the *Herald of Free Enterprise* on a day the vessel was trimmed by the head
- Fire breaking out on the *Scandia* on a lee shore in a gale
- The captain incorrectly measuring distance with the VRM on the *Safari Spirit*
- Engine failure on the *Maria Asumpta* on a lee shore with a foul current
- The assistant captain blacking out at the controls on the *Andrew J. Barberi*
- The position anomaly on the *True North*'s electronic chart

Repetition is mother's milk to complacency; therefore, habitual at-risk behavior may pose the greatest risk to those who *are* experienced. Whereas the novelty of responsibility can keep many greenhorns from becoming overconfident, seasoned mariners, the ones best equipped to see what is wrong with a situation, may not notice what is wrong because they don't examine the situation as closely as they once did. This is not to say that experience always leads directly to complacency. Often the opposite is true: The little precautions and habits of seasoned mariners are often built on hard-learned lessons. When it comes to early detection, experienced mariners have an edge. Still, life supplies far more close shaves than actual consequences, which can persuade people to think more highly than they should of their ability to control events.

The Ladder of Inference

Inferring is arriving at a conclusion based on something else that is held to be true. It is a form of assumption, and assumptions are central to complacency. The *ladder of inference*, a concept developed by Chris Argyris, is a way of thinking about how our brains use assumptions to rapidly sort information and draw conclusions. The first rung of the ladder of inference is where we pick up raw data. Like a first glance at a photograph, there is no meaning, only an image. At the next rung, we begin to attach meaning to the picture, focusing on certain details while suppressing or ignoring others. At the third rung, we interpret the data based on what our individual experiences have shown to be true and real. Then we draw conclusions based on those interpretations and make decisions. The human brain can accomplish all this in an instant. If our inferences and assumptions are correct, the ladder is secure and we benefit from that rapid process. If our inferences are incorrect, then we find ourselves on a flimsy, unsupported ladder that may abruptly collapse. Too fast a climb up the ladder of inference can be hazardous.

One reason we may climb the ladder of inference too quickly is pressure or crisis. Crisis precludes the luxury of serene contemplation at the lower rungs of the ladder. A soldier in combat or a police officer pursuing an armed suspect must sometimes leap to conclusions in order to stay alive. Crisis can bring out the best in people, but the need for action can also lead to incorrect assumptions. In a nautical context, a number of things may cause a vessel to abruptly adopt a precarious list. If a watchkeeper leaps to an incorrect conclusion regarding the cause, the attempt to remedy the situation could make matters worse.

Another reason we might move up the ladder of inference too quickly is complacency. We can't be bothered or we don't see the need to examine our assumptions, so we scoot on up without a thought. Picture a vessel just outside the channel showing an anchor ball. The watchkeeper infers that the vessel must be at anchor because the anchor ball is up. He doesn't notice that there is no cable tending from the bow, and he doesn't bother to track it or mention it to the pilot, captain, or the person coming on watch. He may even be distracted by thoughts of going off watch and getting some food or sleep. But in fact the vessel with the anchor ball has just gotten underway and its crew is too busy securing the deck to take the ball down or change the AIS signature or post a lookout, and now the captain of that vessel is conversing with the pilot and the mate is getting coffee. This has the makings of a dangerous situation. When mariners make assumptions, flawed conclusions can occasionally follow, sometimes with severe consequences.

Assumptions

Assumptions are, or should be, the intelligent application of experience. We employ them because they are expedient and usually correct; it would be a mistake to categorically dismiss them. As with power tools, assumptions can save a lot of time, but occasionally somebody winds up in the hospital. Reasonable assumptions are premised on the belief that things will work the way they are supposed to work and that people will do what they say they will do, or at least they will behave according to predictable norms.

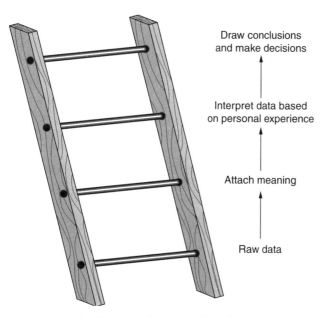

As described by Peter Senge in *The Fifth Discipline Fieldbook*, the ladder of inference is a common mental pathway that has potential to lead to incorrect assumptions.

Generally, mariners assume the following to be true: electronic charts show the vessel where it is, not somewhere else; navaids conform to the chart; crew follow established procedures and will call the captain when in doubt; mariners will not take needless risks. They know these things are not always true, but they are true most of the time, and so crew operate accordingly. Good watchkeeping entails being alert to discrepancies between reasonable assumptions and new information that is coming in all the time. The amount of double-checking watchkeepers do is influenced by what they perceive to be at stake and the amount of effort they feel they can spare. In this sense, distraction, stress, and fatigue can make it harder to overcome or resist complacency.

Assumptions permeate the seafaring profession. For instance, acquiring a license entails passing a Rules of the Road exam with a grade of 90%. This creates an assumption that a licensed individual will remember and apply those rules. There is no refresher course required before going on watch; that would be impractical. When mariners encounter another vessel, all those assumptions are in place for both vessels; yet there is a considerable body of evidence that suggests caution has a place, too. On a case-by-case basis, therefore, watchkeepers refine this grand assumption in various ways: avoiding small CPAs, making Sécurité calls, sounding signals, monitoring targets, and inquiring as to another vessel's intentions and being mentally prepared to take avoiding action. But even with these precautions, basic assumptions remain. Upon agreeing to a port-to-port meeting with another vessel, the two vessels proceed on that basis. If watchkeepers sense a risk of misunderstanding due to a language barrier, a multivessel situation, or some other cause, they may monitor the situation more closely, but otherwise they tend to move on to the next decision. Experience leads them to feel relatively safe in relying on reasonable assumptions concerning the conduct between vessels, and yet, in the next chapter we will see how easily even reasonable assumptions can blow up. When complacency is at work, unreasonable assumptions may cease to stand out from the reasonable ones. Virtually every mishap in this book involved assumptions that appeared reasonable to someone at the time.

Real-World Complacency

Let's recap some of the examples of complacency we've previously discussed. All of these cases directly involved mariners with ample experience who had convinced themselves that nothing would go wrong. Most of them were captains. Consciously or not, they contented themselves with a standard of care that was "good enough" at the wrong time. Sometimes "good enough" really is good enough. Other times, however, it is a warning that complacency is at work.

The *Herald of Free Enterprise*: The working assumption was that the bow doors were closed before sailing. A complacent approach to preparing for sea led deck officers to abandon the written procedures and leave this critical task in the hands of others, although they knew that company vessels had accidentally sailed with the doors open in the past and that there were obvious risks in doing so. Well-trained, knowledgeable people found the arrangement satisfactory, so it persisted, contrary to the basic tenets of seamanship and stability. Part of the tragedy is that not everyone was complacent. Several individuals recognized the risks and urged the company to do something about them but their concerns did not produce results.

The *Andrew J. Barberi*: The assistant captain blacking out in the pilothouse was unexpected, but the fatal accident that followed still could have been avoided. The importance of standard operating procedures regarding the composition of the bridge team had gone unappreciated for a long time. SOPs were erratically disseminated and inconsistently followed and were therefore undependable. This was risky behavior that had become habitual, but it was deemed acceptable on the basis that nothing bad had happened yet.

The *Scandia*: The captain steamed into a well-forecast winter storm with critical equipment compromised, as if it were a routine trip. He was unable to distinguish between the reasonable assumptions and the unreasonable ones. He gave little weight to the risks posed by the weather because many other aspects of the trip were typical.

The *Safari Spirit*: The captain was fully capable of avoiding the rock that cost him his command. But on that particular day he took a minimalist approach to planning his route. He assumed he would not make a simple measuring error, and he was familiar enough with the transit to be content with a minimal plan.

The *Anne Holly*: Here too, overconfidence got in the way of the captain assessing the risks of the situation. If he had suspected there was a chance that he would hit the bridge and jeopardize the lives of over 2,000 people aboard the *Admiral*, would he have done it anyway? Of course not! But complacency rendered that prospect invisible to him.

The *Mercury*: When complacency does take the form of sloppiness, this is what it looks like. Even with simple navigating guidelines to follow, the captain ignored the most rudimentary standards of watch-keeping and went off course. As it is unlikely that this was the first time he ignored procedures, it appears that intentional at-risk behavior may have evolved into habitual.

The *Maria Asumpta*: Though sometimes mariners have warning of mechanical failure, abruptness is a hallmark of such problems. The captain had no reason to doubt his engines; they had been running smoothly earlier that day. But they failed, as engines do, as other things do too. He didn't see the risk in putting the vessel in such a precarious position, and he overestimated his ability to control events.

The *True North*: The captain failed to cross-reference the electronic data with other equipment, especially radar and depth data. Overreliance is complacency in the human-equipment interface.

The Locked Mental Model

In the introduction to Part I, we discussed the significance of a *mental model* that is either held by an individual or shared by a team. A mental model is achieved through standard operating procedures, sound planning, and open communication. It is instrumental in minimizing the unexpected and coordinating the energies of different people. Along with past experience, it helps people compare what is happening to what should be happening. But a mental model is based, at least partly, on reasonable assumptions that can turn out to be incorrect. When the mental model is too rigidly applied, it can hamper the ability to adapt to changed circumstances. As Mark Twain reputedly said, "It ain't what you don't know that gets you in trouble. It's what you know for sure but just ain't so." Whoever said it, it's good advice to ponder. An incorrect or rigidly applied mental model is called a *locked mental model* because situational awareness becomes the prisoner of incorrect assumptions, and it is often complacency that prevents us from breaking out. Both teams and individuals can inherit, develop, and pass along locked mental models, but individuals left to their own devices are perhaps most susceptible to this form of complacency.

The locked mental model is well illustrated by what happened aboard the towboat *Mauvilla* when the pilot became overtasked and disoriented in fog. He had a mental model of where he was and, therefore, what was possible and not possible on that stretch of river. When he saw a well-defined radar echo from a man-made structure extending across the river ahead of him, he was unable to interpret it as anything other than a tow tied off to the riverbank. Because the pilot's mental model was locked, seeing that echo as a bridge was not possible for him. His incorrect interpretation persisted even after the operator of the towboat ahead of him said he had seen no such target just a short time earlier. The pilot twice attempted to make radio contact with the target but received no response. Undeterred, he attempted to raft up with the target without having seen it. Against all other evidence, in his mind the target had to be a tow because there was no bridge in that part of the river; of that he was sure. It never entered his mind that he had turned into a tributary creek. He had climbed a flimsy ladder of inference, and his mental model was locked up tight. Remember, he was trying to do too much at the time; he didn't have the analytical resources to spare so he did the natural thing, which was to shrink his perimeter of situational awareness. A locked mental model is a very human response in some situations as it involves seeing what we want to see, instead of what is there.

It is difficult to defend against doing the wrong thing when you are sure that it is right. For this reason, many of the mechanisms for preventing complacency are useless against a locked mental model. We cannot go through life questioning every apparent fact that presents itself, but we should recognize that the uncritical acceptance of information is what holds a locked mental model together. A healthy suspicion of conventional wisdom and a willingness to look at things a different way can help avoid this. Having a bridge team does not guarantee anything, but if the individuals in the team put forward the effort, it makes it harder for a locked mental model to survive scrutiny.

Deterring Complacency

While we are told to expect the unexpected, it's easier to say such things than to live them. Despite our knowing better, complacency can take root only to be revealed by an unlikely turn of events. The following tactics can help deter complacency and maintain situational awareness.

Planning and procedures: Passage planning, contingency planning, and other forms of risk assessment can be used to highlight foreseeable difficulties and thereby deter complacency around some activities. Standard operating procedures present an opportunity to factor in predictable problems, and periodic training helps remind you to not circumvent those procedures. Standard operating procedures also help defend against the corner-cutting that typifies overfamiliarity. However, if plans and procedures are blindly followed, then complacency will find its way in.

There will always be some procedures that are not written down, ones that you undertake without a checklist. They may be personal habits or vessel-specific practices. Whatever their genesis, they exist for a reason, and, assuming they are safe, they can be defenses against complacency.

Conferring: As long as people are given the opportunity to confer and use it, conferring can be a defense against complacency by exposing an idea to other eyes. Admittedly, the reality of the one-person wheelhouse and minute-to-minute nature of watchkeeping decisions limit the opportunities for conferring for some purposes. Nevertheless, debriefing drills, close calls, incidents, or even a routine success story can elevate awareness of what makes things go well or not go well, and thereby hold complacency at bay. A shipboard culture that encourages all crew members, down to the lowest deckhand, to be analytical and deliberate in the execution of their duties, can help deter complacency and the locked mental model.

New Perspectives: Risky practices are hard to spot once you are accustomed to them. In fact, many bad ideas have been defended on the grounds that they have never caused problems in the past. For this reason an outside pair of eyes can go a long way toward exposing complacency.

Being told by a stranger that a long-standing practice presents a hazard may not be welcome news. But ask anyone who has suffered the severe consequences of complacency if they wish someone had shattered theirs sooner. There is only one answer to that question. The obvious difficulties of self-policing our relationship with complacency has prompted the system of inspections and safety audits now in place. But sometimes the best insights come from the new guy who notices something, and questions it.

Leadership: Complacency can be learned from others; therefore, as with most things, leaders have a role in managing it, and they can set the example either way. If senior personnel are risk-takers or are lax about enforcing standards, their actions will be interpreted as acceptable and will be emulated in turn. For mariners just starting out, that kind of complacency may be especially attractive when it is advertised as the way things are done in the "real world." A culture of complacency, or its opposite, can come to characterize an individual, a vessel, a company, or an entire sector of industry. Leadership has everything to do with this.

*****

Complacency is pervasive and natural. In dealing with it, there are no silver bullets, just a fistful of regular ones. In some cases a blend of simple tactics and strategies may suffice to manage complacency. Other situations may call for deep organizational and cultural changes. Where effective systems and practices are already in place, they should be perfected, not dismantled to make way for something that is new but not necessarily better. Organizations or individual vessels with good safety records have, among other things, usually managed to control complacency.

Whatever combination of traditional or novel techniques are used to deter complacency, they all involve going the extra mile. This requires energy. Like other human factors that degrade situational awareness, complacency doesn't act alone. Mariners who are fatigued, under stress, or exposed to distractions may be disinclined or unable to go the extra mile. Those engaged in highly repetitive work can find it hard to maintain a high level of vigilance. If, as a population, mariners were an especially complacent crowd, there would be a lot more accidents. Working on boats comes with many variables, and this alone tends to moderate complacency. But mariners are also creatures of habit, and that can be a source of complacency. Virtually every accident involving human error contains a kernel of complacency.

Chapter Ten

Transition

THE FOCUS IN RECENT CHAPTERS has been on human factors that diminish situational awareness. Complacency and overreliance are natural human tendencies that lower situational awareness, while distraction, stress, and fatigue are related to the limitations of the human mind and body, which likewise influence situational awareness. The final chapter of this section is on the relationship between situational awareness and transition—transition of any kind.

Transition is not a human factor in the same sense as others we have discussed. It is more of a circumstance that can amplify other factors that lower situational awareness. Transitions have a way of unsettling priorities as we attempt to bridge the gap from the way things were to the way they are becoming. By definition, therefore, periods of transition are more demanding, more stressful, and present more opportunities to become distracted. Transitions are everywhere in vessel operations and mariners must be adept at handling them. If we are interested in being mentally prepared to defend against human error, periods of transition are a good place to concentrate.

The Tug *Seafarer*, the M/V *Balsa 37*, and the Tug *Capt. Fred Bouchard*, 1993

In the predawn darkness of August 10, 1993, the tug *Seafarer* was pushing 236,000 barrels of jet fuel, gasoline, and other "red flag" products past the sea buoy on its way to Tampa, Florida. The mate had the watch, and, since both he and the captain had a first-class pilotage endorsement for Tampa Bay, no local pilot was aboard. Up ahead, also inbound, was the tug *Capt. Fred Bouchard*, pushing 120,000 barrels of heating oil. Due to a mechanical problem with the starboard engine, the *Bouchard* was running on the port engine only. An escort tug (the *Edna St. Phillip*) was made off to the starboard bow of the barge to compensate for the malfunctioning engine. The tow was handling fine, just a little slower than usual. A local pilot was aboard the *Bouchard* tow. Outbound was the foreign flag freighter *Balsa 37*, also with a pilot. Of the three units, the freighter was the smallest and most maneuverable.

The *Seafarer* was making about 8 knots to the *Capt. Fred Bouchard*'s 6 and was slowly creeping up on the *Bouchard*. At 0500, when the *Seafarer* was about a mile astern, the *Bouchard* pilot radioed to ask if the *Seafarer* intended to overtake. Due to traffic considerations, the *Seafarer* mate declined and instead slowed to match the speed of the slower unit. At about 0530, the *Seafarer*'s captain came to the wheelhouse for the change of the watch. The mate briefed the captain on the situation, during which time the mate again spoke with the *Bouchard* pilot and again declined to overtake. Upon taking the watch the captain requested that the mate return to the wheelhouse after breakfast to help with the trip up the channel. As the mate headed below, the captain instructed him to put the throttles to full ahead, which the mate did. Without apprising the *Bouchard* pilot of his intentions, the captain then steered out into the middle of the channel and set about overtaking the *Bouchard* tow.

Meanwhile, on the outbound *Balsa 37*, the pilot told the captain that he could dismiss the lookout because visibility was good. The captain did so, and then he went below too, leaving the third mate on the bridge. At approximately 0535, the chief mate, who had been securing the ship for sea, arrived on the bridge and relieved the third mate. The chief mate monitored the helmsman and transmitted the pilot's commands via the engine order telegraph, but did not take an active role in navigation or monitoring traffic, nor did the pilot ask him to.

At about 0540, the *Bouchard* pilot (the slower tow) and the outbound *Balsa 37* pilot made a port-to-port passing agreement by radio. A moment later, the captain of the inbound *Seafarer* (the faster tow) radioed the *Bouchard* and said, "I'm on your quarter. I'm coming by." This was not a request, but the *Bouchard* pilot agreed to it on the condition that it was okay with the pilot aboard the oncoming *Balsa 37*. The captain of the *Seafarer* then contacted the *Balsa 37* and arranged a port-to-port meeting, without explaining that he was in the process of overtaking the *Bouchard*. The pilot of *Balsa 37*, who had been in a separate radio discussion with his dispatcher regarding his next assignment, had missed the earlier exchange between the *Seafarer* and the *Bouchard* concerning overtaking and therefore was unaware of the maneuver. The *Balsa 37* chief mate, steadfast at the engine order telegraph, later claimed that the radio was turned down too low for him to hear what was being said. If true, it only underscores how detached he was from events transpiring on his bridge, on his watch. Neither the pilot nor the chief mate on the *Balsa 37* bridge noticed the overtaking maneuver on the radar. Additionally, the *Bouchard* pilot, who was aware of the situation, did not clarify the overtaking to his fellow pilot on the *Balsa 37* after the *Seafarer* captain neglected to provide that piece of information.

Thus, the three units converged at a 21-degree bend in the channel in the middle of an overtaking maneuver, in darkness and with no shared mental model. The captain of the tug *Seafarer* and the mate, who was now back from breakfast, watched for the anticipated turn of the *Balsa 37* from Mullet Key Channel into Egmont Channel while checking their progress against the *Bouchard* tow on their starboard side. Somewhere in this interval the *Balsa 37* pilot sighted the green light on the starboard side of the *Seafarer*'s barge and grasped the fact that the *Seafarer* was not behind the *Bouchard* at all but was midchannel and partway through an overtaking maneuver. "What the hell are you doing?" he radioed the *Seafarer*. "You said one whistle!" the reply came back. "I guess we better make it two," the *Balsa 37* pilot responded.

At about 0546, with collision imminent, the *Balsa 37* pilot and the *Seafarer* captain abruptly attempted a starboard-to-starboard meeting at the bend in the channel. Rudders were thrown hard over and engines were ordered full astern. On the *Balsa 37*, however, the chief mate fled the bridge to get the captain so the full astern order was never carried out. The bows met starboard to starboard at a combined speed of close to 20 knots, sending a shower of sparks into the early twilight as the *Balsa 37* dragged down the starboard side of the *Seafarer*'s barge. The *Balsa 37* then careened away from the *Seafarer* and drove directly into the port bow of the *Bouchard* tow, gouging a large hole in a forward cargo tank. The escort tug, which was made up to the starboard bow of the *Bouchard* barge, escaped destruction only because it was the *Bouchard*'s starboard engine that happened to malfunction that day, not the port one.

With the *Balsa 37* flooding in multiple places, the pilot drove the ship aground to prevent it from capsizing. Flames erupted from the jet fuel in the *Seafarer*'s stricken barge, and fire swept toward the wheelhouse. Unable to disengage from the barge, the captain moved to abandon ship. Hoping to put the tow aground, he set the throttles dead slow ahead as heavy black smoke and flames enveloped the wheelhouse,

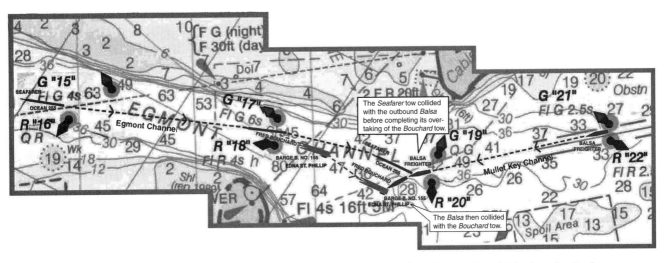

According to the accident report, the inbound *Bouchard* tow was approaching buoys 19 and 20 when the *Seafarer* tow started to overtake. At the same time, the outbound freighter M/V *Balsa 37* was steering a course that would bring it near the center of the channel at the turn into Egmont Channel. The *Seafarer* tow collided with the outbound *Balsa 37* before completing its overtaking of the *Bouchard* tow. The *Balsa 37* then collided with the *Bouchard* tow. (Note, vessels not depicted to scale.) (Courtesy USCGA *Proceedings of the Marine Safety & Security Council*/Ann Aiken)

blowing out the windows. The crew mustered on the aft deck, but they could not reach the life raft. One by one, they went over the side into oily waters topped with fire and smoke. Then a massive explosion blew the top off a cargo tank, sending flames and debris skyward.

The *Seafarer*'s barge was destroyed. The freighter *Balsa 37*, the tug *Seafarer*, and the barge belonging to the *Capt. Fred Bouchard* all suffered severe damage. The cleanup costs alone exceeded $25 million. Miraculously, nobody was killed. Staying behind the *Bouchard* tow would have added half an hour to the *Seafarer*'s voyage.

Failure to Transition

The various versions of just who was where and who was doing what as these vessels intersected have never been fully reconciled, although it appears that the *Balsa 37* was closer to the centerline than it needed to be, aggravating the hazards already posed by the overtaking situation. If the *Balsa 37* had turned earlier, the incident might have been avoided, but the fact that it didn't was the chance aberration required in every case of shattered complacency. However, the undisputed elements of the case leave more than enough fodder for any BRM discussion.

Complacency, distraction, and poor communication are prominent themes, but a high concentration of transitions helped sustain a lack of situational awareness and pave the way toward catastrophe. Two of the vessels had pilots, meaning that the respective bridge teams each had a new member. On the *Balsa 37* the captain and pilot conferred about the transit, but then the captain left the bridge, the chief mate replaced the third mate, and the lookout was dismissed. To the extent a shared mental model was ever created, it certainly did not survive these personnel changes. Aboard the *Seafarer* a change of the watch had just occurred, which resulted in a conservative approach being replaced by a more aggressive approach. A sound mental model appears to have been a casualty of that transition. The overtaking itself was a transition since it altered the traffic pattern from ordinary to more complex.

A channel like the one at Tampa Bay represents a transitional zone between the open sea and the port. Numerous operational and navigational adjustments generally occur in and around such areas. Both the *Balsa 37* and the *Seafarer* were at sea speed, one eager to get in and the other eager to get out. In the case of the *Balsa 37*, the port routine was changing over to the underway routine, which contributed directly to the high turnover on the bridge. The vessels came together where one leg of the channel transitions into the next. This location unnecessarily complicated the three-way meeting with course changes, thus hampering the ability of the watchkeepers to anticipate how that meeting would shape up. Several pilots and local captains later expressed surprise that anyone would try to fit three ships at the same time in any of the bay's channels, let alone at a bend.

Individually, these transitions may be seen as unremarkable, some even routine. But taken together, they contributed to an *accumulation of risk* and low situational awareness on all three bridges. If the people involved had been more alert to how transition impacts situational awareness, they could have managed the risks far better.

Managing Transitions

Personnel changes are a routine and necessary part of vessel operations. While certain efficiencies accrue when people get into a groove and a routine, crew changes are part of working on boats. The resulting transition, however necessary, can create a window of diminished situational awareness. Also, personnel changes come with *personality* changes. While it's tempting to think that personalities do not impact performance in a truly professional setting, in reality they are intensely influential, and they have the power to raise or lower the performance of others.

Changes of the Watch

From a watchkeeping standpoint, there is no transition more important on a daily basis than the change of the watch. The watch handover is a point of vulnerability, as the situational awareness of the off-going watch may be fading while the oncoming watch is still ramping up. Tradition has dealt with this reality through the ritual of relieving officers reporting to the wheelhouse early enough to confer, observe, and absorb the situation before taking over. Even with these measures there is some risk that the oncoming officer(s) won't be fully invested in the watch or fully grasp what is happening right away.

The officer on the *Queen of the North* had just taken the conn a few minutes before missing a course change. Though he went through the motions of taking charge, clearly his mind was not fully on the job. The course change, yet another transition, presented an immediate opportunity to reveal his lack of focus. In the case above, the captain of the *Seafarer*, who had just taken the watch, seems not to have appreciated the traffic situation as comprehensively as the person he had relieved did, or how his decision to overtake could produce a dangerous scenario.

Even when the watch handover is conducted with care, there is a possibility that a complete overlap of situational awareness between the officers may not be achieved. In July 2001 the towboat *Elaine G* was pushing fourteen barges in the Ohio River in moderate fog and predawn darkness when it ran down a recreational vessel, killing all six boaters. The manner in which the recreational vessel was operating had much to do with this tragedy, but the collision coincided with the change of the watch in the pilothouse. Investigators noted that the recreational vessel may have entered a blind spot (visual or radar) of the tow while the attention of the captain and pilot was momentarily diverted to the business of the watch handover. A dedicated lookout could have helped bridge that gap, but there wasn't one.

The watch handover involves the transfer of information, so we must be alert to the possibility of a communication breakdown. Some operators use checklists to facilitate the watch handover, and some officers jot down notes in the course of a watch. While it seems unthinkable that an officer would neglect to confirm the vessel's position and the course, study the radar picture, and discuss the traffic picture at the watch handover, such omissions have occured. And if a more subtle but significant matter has occurred hours earlier (such as a mechanical issue, a radio broadcast, or a verbal reminder), it's even less surprising if it doesn't get passed along properly. For this reason checklists and written notes can be useful prompts in the watch transition, but like other documentation, they are not helpful if they become a distraction or an exercise in generating paper trails that satisfies the legal department without benefiting the watch officer.

Even with memory prompts, a relieving watchkeeper must work to develop his or her own mental model if only because the off-going watchkeeper is fallible and may have missed a development. Classic cases of change-of-watch oversights include a new danger target emerging from the edge of the radar screen, a weak target hidden in sea clutter or a blind spot, or a target vessel that has just altered course or speed, thereby transforming the traffic picture. Indeed, watchkeepers need to be looking for these things at all times, but when they occur as the watch is being handed off, they may be especially easy to miss. Another reason the oncoming watchkeeper must make a conscious effort to establish situational awareness is the possibility that a locked mental model, with all its assumptions and misconceptions, can be transferred intact from one person to another. The only defense against this kind of transitional hazard is for relieving watchkeepers to do their own homework: check the status of equipment, scan the horizon, shift range scales. You can take the information offered, but confirm it and draw your own conclusions.

Finally, the handoff should be clearly stated ("I got it"), and the transfer of responsibility should be transparent to all involved so that new information is channeled where it needs to go. It is important to note that this need for clarity also applies when the captain takes charge for any reason. There have been numerous occasions where a captain has begun to direct the vessel incrementally or intermittently without clarifying whether the officer still has the conn or not. In some circles this behavior may be perceived as a privilege of the master, but the practical hazards of this type of transition, not to mention the legal ramifications, are obvious.

Crew and Leadership Changes

Crew changes and changes of command occur less frequently, but they are significant transitions nevertheless. Aboard the *Herald of Free Enterprise* the practice of moving officers between different vessels and different runs meant that some personnel were in a permanent state of transition, and situational awareness suffered.

In another example, in 2007, a new anchor-handling vessel, the *Bourbon Dolphin*, capsized in the North Sea while moving an oil rig, claiming the lives of the captain, his fourteen-year-old son, and six others. The incident raised a host of technical and procedural issues related to the complexities of the operation but among them was the fact that it was the first trip on that vessel for both the captain and the first officer. They were unfamiliar with the vessel, the crew, and the task. The captain had ample anchor-handling experience with a bigger vessel, but it was precisely that background that may have led him to underestimate the challenges of the smaller vessel. The first officer, who had been with the vessel a few months, had experience with offshore supply vessels but minimal anchor-handling experience, and none of it in deep water. The brevity of the handover with the off-going officers and the lack of experience with the new vessel itself exacerbated the highly transitional nature of this particular situation, and created greater risks than the operators realized.

Integrating Pilots

Many small-vessel operations do not ordinarily use pilots, but boarding a pilot is a special personnel change that warrants its own discussion. Pilots are technically advisers to the captain, either directly or indirectly through a watch officer, but their level of involvement makes their role far more significant than that description implies. A pilot is a newcomer who immediately begins to direct the movements of the ship, yet the captain and crew retain responsibility for the outcome. If the captain goes below, leaving an intimidated or unengaged mate on the bridge, the arrival of a pilot can amount to a change of command and the loss of a bridge team. This is not the way it is supposed to work, but it often has, and this is what happened aboard the *Balsa 37*. Consider that 90% of all ship casualties occur in confined waters, and that 60% of those occur with a pilot on board. This isn't because pilots are bad navigators; it is because the entire situation is exposed to the full potential of human error. The loss of situational awareness in the wheelhouse with the arrival of a pilot can occur in several ways, including the following:

- Overfamiliarity with waters, resulting in complacency on the part of the pilot.
- The presence of the pilot, and his or her implied expertise, can breed complacency in other members of the bridge team, who may relax or become order-takers at a time when the risk of danger is at its highest. This happened on the *Balsa 37*.
- The short-term nature of the pilot's integration can frustrate the development of a shared mental model, particularly if it is disjointed, as on the *Balsa 37*.
- The unrehearsed working relationship between a ship's crew and a pilot can lead to communication breakdowns and misunderstandings under the best of circumstances. Language and cultural differences can present additional barriers to good communication and cooperation. There is reason to believe that communication and cooperation were not optimal on the *Balsa 37*.
- In addition to the inherent risks of shallow water and traffic, pilotage waters often come with more distractions and stress to the detriment of situational awareness among all the participants. Shortly before colliding, the pilot on the *Balsa 37* was talking to his dispatcher and missed the critical radio exchange in which the other two vessels discussed overtaking.
- Pilots suffer from fatigue too, but the transitory nature of their participation may make it difficult for ship's crew to detect problems with fitness for duty.
- The pilot's relative unfamiliarity with the vessel, its equipment, and handling can lead to a variety of miscalculations and errors.

Pilot integration is recognized as a transition of great significance requiring its own procedures, and as a result the master-pilot exchange (MPX) is integral to BRM (see Appendix). We will look more closely at those procedures in Chapter 12, Teams and Teamwork.

Organizational Changes

Mariners are well aware that changes ashore can result in changes aboard. Organizational changes can happen in response to many things: new management, policy or regulatory requirements, industry restructuring, customer needs, etc. In the short term, the process of adjusting to the new procedures can distract from well-worn fundamentals and what was best about the old way. Organizational changes, like other transitions, can leave crew vulnerable if the changes are not integrated carefully.

Recall that the management behind the *Herald of Free Enterprise* was having difficulty getting its cross-channel ferries out of Dover, England, on schedule. This was not due to some specific inefficiency; there just wasn't enough time in the schedule. Rather than allowing more time, management attempted to "create" time by getting the vessel out of Zeebrugge ahead of schedule: "put pressure on the first officer if you don't think he's moving fast enough.... Let's put the record straight, sailing late out of Zeebrugge isn't on. It's 15 minutes early for us." The quest for efficiency comes with most businesses, and most means of achieving it are not as ill-conceived as that one was. Change cannot, nor should it, be avoided simply because it entails transition. But organizational changes can interfere with normal levels of vigilance at the operational level, and maintaining situational awareness may require an above-average effort during such times.

Changes of Plan

We have addressed the importance of planning as well as the importance of being able to adapt to new circumstances. Both are part of watchkeeping, but remember that a change of plan is also a transition. As you let go of the old plan and begin to follow a new one, a drop in situational awareness is a possible consequence. This happened with the *Mauvilla* when fog caused the pilot to stop trying to get where he was going and start looking for a place to tie up. This was a sensible change of plan, but it contributed to his loss of situational awareness when he ended up doing a bit of both, and neither one well. On the *Maria Asumpta* the captain incrementally changed his plan until he unwittingly surrendered all margin of safety. In the Tampa Bay case discussed above, there were two changes of plan. The captain of the *Seafarer* decided to overtake the *Bouchard* after the mate had told the *Bouchard* that the *Seafarer* would not overtake. By the time the *Balsa 37* and the *Seafarer* attempted a starboard-to-starboard meeting, another change of plan, stress levels were doubtless very high and good judgment departed. Collision was probably inevitable at this point, but the change of plan had little chance of success and helped turn a two-way collision into a three-way collision. Sometimes the plan has to change, but when it does you need to make sure you are replacing it with something viable and that you have the resources to pull it off.

When a change of plan involves the coordination of other people on other vessels, as in the Tampa Bay incident, the transition must be handled even more deliberately so as to create a new shared mental model. This usually means more lead time. If more lead time is not available, then maybe it's not the right plan. Plans are important, but so is flexibility. Changing the plan on short notice is a transition that can have unintended consequences.

Operational Transitions

Vessel operations frequently entail transitions that are related to the fact that mariners operate in changing conditions, around the clock. These transitions can influence situational awareness.

Twilight

Hidden in the word *twilight* is the word *two*, conveying the way in which twice a day there is neither day nor night but both at the same time. For a watchkeeper, going from night into day is normally a welcome transition because everything is easier when you can see where you are going. Nevertheless, the gray light of dawn is a time when depth and color perception are weak, making it difficult to pick out objects and lights. Due to circadian rhythms, levels of alertness are near their lowest ebb in the period just before dawn, which is when the three-way collision in Tampa Bay occurred.

The transition from day into night entails the loss of light and is therefore a more demanding transition, prompting greater reliance on instruments. In either case, twilight can find a watchkeeper overrelying on a given resource or method that was more appropriate a few minutes earlier than it is to the situation unfolding now. Moving between different sources of information is part of good watchkeeping, but it takes on special significance when transitioning from day into night and night into day.

Visibility

The transition from good visibility to restricted visibility poses similar problems as twilight. It is easier to make the change from restricted visibility to clear conditions; yet more than one mariner has been caught with his or her eyes still glued to the radar or chartplotter when looking around would have served better. The pilot of the *Mauvilla* faced the more difficult challenge of going from clear visibility into fog. He attempted to respond to the new circumstances but could not bridge the gap from one situation to the next. Operating in fog for long periods is draining, in no small part due to the persistent uncertainty. Mariners can and do adapt, but the shift is not necessarily instantaneous.

In July 2008 the U.S. Coast Guard cutter *Morro Bay* and the ferry *Block Island* banged into each other in Block Island Sound when fog abruptly shut in. Neither bridge team was aware of the presence of the other vessel as both were in the process of making "operational adjustments" with regard to speed, lookout, use of radar, and standing orders. The captain of the Coast Guard cutter was summoned as his bridge team shifted to "restricted visibility mode," but events were a step ahead. The fact that it was the captain's first day on the job was a transition that may or may not have been a factor, but it is clear that the bridge teams on both vessels were unable keep pace with changes in their operating environment.

Bottlenecks

Bottlenecks are transitional zones that funnel vessels between bodies of relatively open water or in and out of harbors, bays, and sounds. Bottlenecks require elevated situational awareness for obvious reasons: traffic density, physical constraints, and currents. Normally our situational awareness rises to the occasion, but we should not take this shift of mental posture for granted. If, due to other human factors, situational awareness does not match the new situation, then these transitional zones will be trouble. If the transit involves unfamiliar waters, yet another layer of transition is at work as you attempt to assemble local knowledge on the spot.

Way Points and Waterways

The transition from one heading to another at a way point or a change in the orientation of the channel is a seemingly minor affair, yet turning too late or too early can have serious consequences. Changes in the orientation of a channel entail a change in perspective, and that can be disorienting, especially in unfamiliar waters or with the low height of eye that many smaller vessels have. Three experienced bridge teams in the Tampa Bay collision failed to give this fact much consideration. When a turn is coupled with a meeting situation, let alone overtaking *and* meeting, the inherent hazards of this particular transition are multiplied. Throw in some fog, darkness, or distraction, and a course change, an act of basic seamanship, can represent a significant elevation of risk.

The container ship *Ever Grade* and the Coast Guard buoy tender *Cowslip* found this out in May 1997 when they agreed to meet in fog near a bend in the Columbia River in Oregon. Meeting at the bend magnified the effect of other operational deficiencies, and the collision that followed pinned the *Cowslip*'s throttle in the full ahead position, nearly crushing the people on the bridge. Damages to both vessels came to $1.2 million. It was within the control of the two vessels to meet a little upriver or a little downriver, and if they had it would likely have been a nonevent.

Getting Underway

There's an old saying that port rots ships and sailors. Setting aside the traditional explanation for that maxim, a bridge team (including a team-of-one) that has just gotten underway may still be in the process of discarding the in-port frame of mind and replacing it with the underway one. For any number of reasons (paperwork, final calculations, stowing for sea, maneuvering, communicating with the engine room) focus may be fragmented and situational awareness blunted. Though it's difficult to quantify the role of this kind of transition, we should note that the *Safari Spirit*, the *Anne Holly*, the *Herald of Free Enterprise*, and the *Balsa 37* all got into trouble shortly after getting underway.

Changes Related to Bridge Equipment

The point of most bridge equipment is to raise situational awareness by providing alternate perspectives and multiple sources of information. As we know, misuse of these devices can be worse than no devices at all, as without them we are likely to be more conservative. The use of bridge equipment comes with its own transitional moments that can lower a watchkeeper's situational awareness if he or she is not alert to the opportunities for this to happen. Here are a few examples.

Chart Changes

The transition from one chart to another has always presented opportunities for a navigator to make mistakes. If the charts use different scales, it is easy to misinterpret the latitude and longitude markings on the edge of the chart, leading to incorrect plots and distance measurements. Changes in chart datum, sounding units, the shading of bottom contours, and other details from chart to chart may not register, especially if you are in a hurry. This can lead to false assumptions concerning the displayed data.

Electronic charts sidestep some of these issues, but present others. Because the level of detail displayed on a vector chart depends on the scale, it is possible to miss details by not zooming in enough, or to lose track of the big picture when zooming in for more detail. While all the relevant information is available, the interrelationship between scale and detail can lead to confusion when moving between different perspectives. Something similar can happen with radar, where detail is sacrificed for greater range, and vice versa, or suppressions are used to manipulate the picture.

Changes in Settings

Just about every piece of bridge equipment has settings and adjustments that can end up inappropriately set for a given situation. For instance, VHF radio lets you communicate with and monitor the communications of others. However, if you shift channels or lower the volume for valid reasons and fail to reestablish the appropriate settings, you will be out of contact with those around you and not even know it. Low volume allegedly contributed to the chief mate on the *Balsa 37* being out of the loop regarding the traffic situation. One seasoned mariner summed it up this way: "Every watch I start on the left of the electronics array and adjust everything, even if it's not needed. Amazing what you find that way." You should check bridge equipment several times per watch.

The notion employed by some captains that only one person can touch the knobs can be counterproductive in some situations, but when more than one person is using bridge equipment at a time, there is a clear opportunity for settings to be changed, resulting in misinterpretations and errors. In such situations, communicate your intentions with regard to the equipment, such as, "I'm bumping down the range scale, okay?"

Changes in Equipment

Even when better equipment comes along, many mariners prefer units they are familiar with. That preference is not mere nostalgia; knowing what the buttons do means mariners are more likely to use the equipment correctly. The installation of new equipment, therefore, entails a transition that can temporarily decrease competence and situational awareness while crew learn the new "knobology." Aboard the *Queen of the North* the deckhand's lack of familiarity with the new autopilot rendered her unable to respond to the crisis by swiftly switching to hand steering. Many, many moments of confusion have arisen around the installation of new bridge equipment, usually because time for proper orientation has not been made.

On balance, the opportunity afforded by bridge equipment to change perspectives and gather new information is beneficial. However, these capabilities come with a risk that watchkeepers will adopt the wrong perspective or fail to maximize the capabilities without realizing it. In a bridge team, rational people can disagree on what constitutes the best perspective or the most important information, but good watchkeeping always involves constantly confirming that you are using the optimum perspective for the situation, that you know what you are looking at, and that you know what the buttons do so you are not in the dark.

*****

Transitions are all around us, all the time. They keep life interesting. But as watchkeepers we should be aware that transitions are times when a person's head may still be in the old game after a new game has begun. A transition is a built-in distraction that leaves us spread more thinly than at other times. Routine allows us to conserve energy and consolidate focus; transition has the opposite effect. Remembering this allows us to adopt strategies that will leave us less exposed to a loss of situational awareness.

Part III

Human Interaction

THIS IS A BOOK ABOUT PEOPLE, not boats. The only process that happens on a boat without people is rust. The overriding premise of BRM is that it is not enough to be a capable navigator, boat handler, or marlinspike sailor, or to be knowledgeable in the intricacies of a given cargo. These, and many other skills, are all part of what make a complete mariner. We have seen, however, that human factors, both individually and in groups, are also integral to success and failure. In our discussion of BRM principles we have used watchkeeping as the home base, circling back to it for the main points. This focus is appropriate but it should not limit our ability to project BRM ideas into other aspects of shipboard operations, and into organizational behavior in general.

Part I examined how planning and procedures can improve performance by clarifying expectations and minimizing the unexpected. "Control the controllable" is the essence of this aspect of watchkeeping. Part II demonstrated that plans and procedures only go so far. Situational awareness is essential for executing and monitoring those plans and procedures, and adapting to change. But situational awareness is a dynamic resource that ebbs and flows with other human factors and cannot always be relied upon. Because situational awareness is related to the natural limitations of human performance, it cannot be turned up like the wick of a lantern. We can, however, be aware of those things that degrade situational awareness and work to manage them.

The issues discussed thus far have raised self-awareness concerning individual professional performance. If that were all we had to think about, it would be enough. But the operation of a vessel, even with a team-of-one watch arrangement, invariably requires interaction and cooperation with other people who are aboard, ashore, or on another vessel. With other people involved, the job can get easier in some respects but more complex in others. Many hands *do* make light work, but now you also have to ask, explain, reiterate, follow up and, yes, listen. This final section looks at human interaction in the shipboard environments, focusing on communication, teamwork, decision-making, and leadership. These topics are obviously relevant to the running of ships, and always have been. That the 2010 Manila Amendments to the STCW Convention singled them out for new training competencies merely recognizes that such things should no longer be left to chance or osmosis. Also in this section we will examine certain patterns around the phenomenon of human error and explore some ways of defending against it.

Chapter Eleven

Communication

Docking a large ship is an immensely impressive yet routine operation: the largest mobile structures on the planet are spun around in their own length and tucked neatly into a parking space with a few feet to spare. The juxtaposition of no-nonsense functionality and the nuanced use of power and vectors is one of the more incongruous displays of muscle and grace on today's waterfront. You need not participate in the multivessel waltz to know that good communication is critical. With such substantial forces at work, a minor misunderstanding can result in severe damage or injury. Yet mariners also know that good communication doesn't need to be elaborate to be effective. In the nautical world successful communications might well be summarized as the simpler, the better. But keeping things simple is not the same as saying nothing, or too little. Silence is not golden in the management of vessels.

Communication has always been integral to human interaction, but it has a special place in the maritime world. The distinctive terminology, the global nature of the work, and the inherent challenges of communicating within a ship, between ships, and with those ashore have made for a unique communications laboratory down through the centuries. Archaic as flag, sound, and light signals may seem in an age of instant messaging, the ingenuity of those systems reflects the supreme significance, and the elusiveness, of effective communication in vessel operations. The ancient practice of repeating commands and acknowledging them was never about hierarchy; it was about preventing misunderstandings. For good reason, these tried-and-true protocols are with us still, but our options are greatly expanded. As with navigation, advanced communication technology has its benefits, but it doesn't guarantee that good communication will occur. As technology improves, our communication failures may have less to do with the limitations of equipment and more to do with those who use it.

Confining the discussion of communication to a single chapter in a BRM book is perhaps in itself inviting a communication failure; there is much to say on the subject. But in reality, the topic has been with us all along at the margins of other discussions. Discussing communication is like wrestling a giant squid with all its tentacles flailing at once. Is communication about writing, speaking, signals, and symbols? Is it equipment, language, and terminology? Is it style, context, tone, local usages, accent, a rolling of the eyes? It is all of these, and more. Depending on how and where a mariner comes to the profession, certain aspects of communication are more pertinent than others. The challenge professional mariners face is that effective communication requires them to be competent in a complex area that never receives the same emphasis as, say, navigation, cargo, or shiphandling. By breaking down the subject we can better see that it has different components, and that good communication can be learned, improved, and even mastered, like any other skill.

Failure to Communicate: The *Seafarer*, the *Balsa 37*, and the *Capt. Fred Bouchard*, Revisited

In the last chapter we examined a three-way collision in Tampa, Florida, that severely damaged four vessels. (See diagram on page 101.) The events leading up to the accident involved a number of transitions that contributed to low situational awareness. Yet it was evident that poor communication also played a large role. Why didn't people speak up? What could they have said? What could they have done differently? Most important, what would *you* do differently in a similar situation? Let's review the communication breakdowns that occurred that morning in Tampa Bay.

Communication aboard the *Seafarer*: The captain of the *Seafarer*, the faster inbound tug and barge unit, was in the wheelhouse when his mate told the

Capt. Fred Bouchard (the slower inbound tug and barge unit) for the second time that he would *not* be overtaking. Yet with the change of the watch, only minutes later the captain began to overtake without notifying the *Bouchard* of the change of plan. This was a major communication failure that raises other questions about the quality of communication during the watch handover itself. Though the men stood only a few feet apart, they had different mental models and did not reconcile them.

Communication between the *Seafarer* and the *Bouchard*: Eventually the *Seafarer*'s captain did inform the *Bouchard* that he was overtaking, but only after he was on the slower vessel's quarter and about to pass. In communication, not only is clarity important but so is timing, and in this case the cart was in front of the horse. By maneuvering first and talking later, the *Seafarer*'s captain generated momentum and pushed caution aside. Nor did the manner of his communication ("I'm on your quarter. I'm coming by.") leave much room for consultation. Momentum drove events, rather than sound seamanship. The lack of proactive communication helped elevate the maneuver to the status of a "done deal."

The *Bouchard* pilot (or the captain, who was also on duty) could have initiated contact with the *Seafarer* once the overtaking had commenced to request clarification of the *Seafarer*'s intentions and to raise the issue of the outbound *Balsa 37*, whose course appeared likely to converge with theirs at a bend in the channel. There was time to make this contact, and the pilot was not overtasked. The *Bouchard* pilot had a good reason to seek clarification because the *Seafarer*'s actions contradicted what had been agreed to just moments earlier with the mate. But he didn't.

When the *Seafarer* finally did contact the *Bouchard*, the *Bouchard* pilot had the option of recommending that the *Seafarer* hold off because of the outbound freighter, *Balsa 37*. Since the overtaking maneuver was in an advanced stage by this time, such a response would have meant challenging the *Seafarer* captain's intentions over the radio. This is never a first choice, yet effective communication may sometimes require it. Instead, the *Bouchard* pilot put his fate into the hands of others by saying that it was okay with him if it was okay with the *Balsa 37*. The *Bouchard* pilot viewed his presence as peripheral but it was not. One lesson here is that if

something is happening near you, then it may end up happening *to* you.

Communication between the *Seafarer* and the *Balsa 37*: After the exchange with the *Bouchard* pilot, the captain of the *Seafarer* contacted the *Balsa 37* to propose a port-to-port meeting. In the greatest communication failure of all, the captain neglected to mention that he was no longer following the *Bouchard* tow, but was in the act of overtaking it. Perhaps he assumed that the pilot on the *Balsa 37* had overheard his intention to overtake the *Bouchard*. Since the *Seafarer* captain's actions did the most to alter and complicate the traffic picture, there was a responsibility to not only consider the potential ramifications but to ensure there was a unified mental model among all three bridges. This was not accomplished.

Communication aboard the *Balsa 37*: Around the same time that the *Seafarer*'s captain contacted the *Bouchard* about overtaking, the *Balsa 37* pilot was engaged in a separate radio call regarding his next assignment. Though business related, it distracted him from his primary duties, and he did not overhear the *Seafarer*'s captain tell the *Bouchard* pilot that he intended to overtake. It's easy to miss the occasional transmission amid the constant buzz of relevant and irrelevant radio chatter on a busy waterway. An effective bridge team can help backstop oversights like that. But the people on *Balsa 37*'s bridge were not a team, and they all missed it.

The lack of effective communication on the bridge of *Balsa 37* became apparent again when, as the vessel approached the bend in the channel, the pilot ordered a course change from 261 degrees to 262 degrees that was never carried out. The mate and the helmsman later said no such order was given. Too many other things were already going wrong to blame the collision on this breakdown, but it did mean that the *Balsa 37* ended up nearer to midchannel than it needed to be, or the pilot meant it to be.

Communication between the *Bouchard* and the *Balsa 37*: One of the chief benefits of carrying a pilot is that, in addition to knowing the waters, the pilots know each other. That familiarity can help them communicate more effectively than strangers might, especially when local usages (and languages) differ from those prevailing aboard. The pilots aboard the *Bouchard* and the *Balsa 37* were colleagues. They made a port-to-port passing agreement but neither referred to the fact they would be meeting at a bend in the channel. This was a missed opportunity that complacency may have played a part in.

The *Seafarer* captain's failure to clarify his overtaking maneuver to the *Balsa 37* pilot was a crucial piece of situational awareness that the *Balsa 37* pilot lacked. Yet the *Bouchard* pilot made no effort to fill that gap for him. There was no clear requirement to do so, and perhaps he too assumed that the *Balsa 37* pilot had overheard the exchange. Greater appreciation for the infidelity of assumptions might have helped him to see that this part of the developing situation should not be left to chance. The failure to make a radio call at that time left the *Balsa 37* pilot out of the loop at a critical time.

The *Seafarer*'s attempt to overtake unnecessarily courted risk. What made a bad idea even worse was complacency, distraction, and a collective failure of the captains, mates, and pilots on all three vessels to say what needed to be said. Collectively they failed to communicate their thoughts and intentions, or to solicit or challenge the thoughts and intentions of others. For communication to succeed, people have to be interested in communicating. This incident demonstrates what can happen if they are not. While there is such a thing as too much communication, this was a case of too little.

Lines of Communication

Communication can be described as converting internal thoughts to external messages that are understood by the recipient(s). No one method of communication is perfect for every situation; indeed, no one method is ever perfect at all. This is perhaps the most important thing to recognize about communication: failure is easy to achieve, and success should never be taken for granted. In the nautical world, where the potential consequences of communication failures are high, the expectation is that mariners will communicate in ways that will bring the greatest chance for success, while being alert to the possibility that misunderstandings may still arise at any moment. Let's start our discussion with some fundamentals.

The main components of a communication system are a sender, a receiver, a message, and a medium. The sender and the receiver are the people involved. They may be standing side by side holding a conversation, or they may be nameless, faceless, and thousands of miles apart, connected only by the written word. The message is the thing being communicated. It may be a command, an expectation, a piece of factual information, a question, an opinion, a request, an emotion, a hunch, a recommendation, and so forth. The medium is how the message is transmitted. Speech, writing, and signals are examples. In general, the more senders and receivers have in common, the more likely it is that their communication will succeed. Tug masters often make the best docking pilots, not only because they understand ship docking but because, having run tugs, they also know how best to communicate their needs to the tugs they are orchestrating. The less senders and receivers have in common, the more likely they are to misunderstand each other. There was no lack of common ground in the Tampa Bay situation, which only proves how easy it is for communication to break down, even under favorable circumstances.

Marine communication is sometimes classified as internal or external. Internal communication occurs within the vessel, including the watch handover, bridge team interactions, conferring, hollering up and down the deck, and hand signals. External communication is everything else, including radio exchanges with traffic and VTS; whistle, flag and light signals; company memos; email; and calls from the home office. Both internal and external have special challenges.

Face-to-Face Communication

In terms of making ourselves understood, face-to-face verbal communication is the gold standard. While words are the currency of verbal communication, they convey only a portion of our meaning. This becomes evident in a face-to-face setting where the sender can enhance his or her meaning with inflection, intonation, and body language, and the receiver can react in kind. Eyes have been called "the windows of the soul" because they have the capacity to animate language and convey the full spectrum of human emotion. The descriptive qualities of hand gestures are well known and also play a part in face-to-face communication. We often send and respond to these message enhancements unconsciously, but the degree to which they enrich the exchange is incalculable. In a face-to-face exchange, the sender and the receiver rapidly alternate roles, so both parties are sending, receiving, and reacting to nonverbal cues, all of which can make the exchange more efficient and more complete.

Whether we know it or not, we are all amateur lip readers. Being able to see a person's face can help overcome difficulties related to enunciation and pronunciation, especially where similar-sounding words are involved or there is background interference. Turning to face someone when communicating is not only a courtesy, it facilitates the message. However, even in face-to-face settings, environmental conditions aboard ship may present obstacles: squawking radios, interruptions, machinery, wind. Face-to-face communication may also be impaired if gestures and body language are inconsistent with other parts of the message. For instance, if you habitually use your right hand to indicate the starboard side of a vessel, but do so when facing aft, your words will contradict your gesture. There is no guarantee that the receiver will pick up on that contradiction, so it is important to align gestures with the message. No communication mode is perfect but, on balance, face-to-face communication has the broadest spectrum. Knowing this can help us decide when face-to-face communication should be made a priority and when other methods will suffice.

Non-Face-to-Face Spoken Communication

Face-to-face verbal communication is a luxury mariners don't always have. When someone is up a mast or out on a barge, even if he or she is visible, the nuances of face-to-face interaction are lost. The same goes for a conversation with someone who is down in the engine room, or on another vessel, on the dock, or in the office ashore. Information is exchanged but without the benefit of visual cues. This extra degree of separation is a barrier that requires extra effort to overcome. Professional mariners do this by speaking with precision, using standard language, enunciating clearly, making sure that words don't run together, listening carefully, and confirming the message at intervals ("Did you copy that?").

Non-face-to-face spoken communication often relies on devices like radios, cell phones, sound-powered phones, etc. While inferior to face-to-face communication, these devices do allow for at least some vocal inflection and emphasis. They also preserve the ability to seek clarification, state things a different way, or repeat as needed. However, the equipment itself can interfere with the message if people fail to key the microphone properly, the batteries die, wind distorts the message, and other technical problems arise.

For example, in the moments after the *Andrew J. Barberi* slammed into the Staten Island pier, the chief engineer attempted to contact the pilothouse by sound-powered phone to find out what had happened. He was able to make the phone ring on the bridge, but he couldn't hear anyone speaking. Moments later, he received a call from the bridge, but when he picked up, no one was there. Postaccident tests showed that the bridge and engine room could ring each other, but it was not possible to hear what was being said. A device intended to facilitate communication ended up being worse than nothing because it didn't work properly when needed; rather it became a distraction during an emergency and compounded an already stressful situation. While the sound-powered phone did not contribute to the accident, its failure, and the failure to test the equipment systematically, highlights the low priority the organization placed on effective internal communication.

Non-face-to-face spoken communication is an important and necessary element among communication choices, but it has limitations. When those limitations are further burdened by equipment that does not function as intended, serious communication breakdowns can occur.

Written Communication

According to one study, people remember 29% of what they see, 40% of what they see and hear, and 70% of what they see, hear, and do. These figures do little to recommend the written word as the communication mode of choice. The written word obviously lacks the visual and auditory cues that lend power to spoken communication. The ability to engage in immediate give-and-take, request feedback or clarification, and ask questions is also burdened. Written communication can be a two-way correspondence, but it is clearly more cumbersome than speaking. When written communication takes the form of instruction manuals, policies, *Local Notices to Mariners*, and codes of regulations, it is essentially a one-way street. For the sender, who is relatively remote, there is no way to know if the message got through, if it was understood as intended, or if it was even read.

Conversation is dynamic. You can clarify, retract, and adjust as you go. Writing is more rigid. Just as non-face-to-face spoken communication requires extra effort, written communication requires even more care to accurately convey your meaning. The time involved with written communication is another drawback, but when people don't take enough time and instead dash off carelessly crafted messages it can lead to major misunderstandings and unintended consequences. Email has a particular reputation for generating misunderstandings, and even damaging relationships in the process. The informality with which many people approach email seems to lend itself to misinterpretation: attempts at humor backfire; business-like messages come off as arrogant; abbreviations, misspellings, and incomplete sentences can be interpreted as laziness or a lack of maturity, education, intelligence, or respect, though none of these things may be true. The bottom line is that writing is more work than speaking, and the results can be less certain.

For all these disadvantages, the written word can do things that no other mode can, when handled well. Speaking is ephemeral, whereas writing is permanent. Just as the written word would be painfully slow and unresponsive for coordinating a man-overboard response, messages that are expected to endure, such as operating procedures, regulations, and so forth, should not be transmitted by word of mouth. Imagine if the captain had to be awakened repeatedly to recite the standing orders. For some things, writing is better.

In contrast to the game of telephone, once something is in writing, the wording does not change. As long as the writing is precise and unambiguous, it becomes a resource that can be consulted whenever necessary. Writing, though imperfect, is an agent of standardization. For this reason, if no other, written communication will always be with us. In addition to the above benefits, some written communication, such as logbooks, provide a record, which has value for many reasons.

Despite the relative permanence of written communications, they are still subject to interpretation. Aboard the *Herald of Free Enterprise* the written instruction that required the loading officer to ensure the bow doors were secure gradually came to mean that the loading officer would merely confirm that someone was at the station ready to close the doors. Inevitably, the day came when the weakness of that interpretation was revealed. On the *Andrew J. Barberi* written procedures did not produce a standard approach to bridge team composition and practices. In all but the most lucid and simply crafted writing, there is usually room for two people to see different things. This possibility must be borne in mind when relying upon written communication.

Nonverbal Communication

Words are the most natural way humans communicate, but a lot of marine communication relies upon other modes. The purpose of any nonverbal system is to make communication possible when words are unreliable, impractical or, frankly, unnecessary. As mentioned, nonverbal cues such as body language, gestures, tone, and facial expressions can "speak" for us, but in the nautical world a great deal of nonverbal communication has been formalized into international systems of sound, light, flag, and even hand signals. Chart symbols and buoyage systems are also forms of nonverbal communication. These systems are standardized through publications such as the COLREGS, the International Code of Signals, *Chart No. 1*, the IALA-A and IALA-B systems, and so forth. Nonverbal communication systems, official or otherwise, come with an obligation to know the system in use, and to use it systematically. Like language itself though, there are regional, local, and even vessel-specific usages, and the so-called standard systems may not be standard at all.

We have all had the experience of words getting in the way of saying what we mean. While many things are too complex to convey with a symbol or a signal, mariners should appreciate that nonverbal communication is often less prone to ambiguity than words.

While technology has diminished the role of nonverbal communication since the days of Nelson at Trafalgar, mariners need to recognize that words can unnecessarily complicate matters when a signal, or simply complying with rules or accepted norms, will suffice. This is especially true when multiple languages, second languages, or dialects are in play. When operating in a specific port or region, it is both comforting and necessary to know the local vernacular. But ships move; not everyone can be local. If the person you are attempting to communicate with is unfamiliar with the area, then local usages could be an impediment to successful communication. There are times when a more universal approach may be necessary. We should recall that, in the days before VHF radio, cell phones, and the like, the wheelhouse was a far quieter and less distracting place. Sailors made themselves understood by maximizing their nonverbal communication options. When you are the give-way vessel in a crossing situation, turning to show your port bow to the stand-on vessel sends a clear message that can eliminate a world of uncertainty.

True, nonverbal systems do not have the full descriptive and expressive force of language, so they tend to work best for simple messages organized around a single activity or interaction. Another limitation is that some nonverbal systems are affected by environmental conditions; you need visibility to confirm a light, flag, or buoy, and a sound signal must be audible. Despite certain advantages, in some situations nonverbal systems simply won't work or won't convey the appropriate level of detail.

The VHF (and Text) Assisted Collision

Some mariners argue against complicating vessel interactions with verbal communication when the Rules of the Road can suffice. The theory is that talking invites misunderstanding and delays action. When the *Hyuandi Dominion* and the *Sky Hope* collided in 2004 in the East China Sea, avoiding action was not taken until the vessels were only 0.2 mile apart. Instead of getting out of one another's way, the watchkeeper on each vessel squandered precious time on fruitless VHF communication and, astonishingly, a last-ditch attempt by one officer to text the other via AIS.

In the Tampa Bay incident, the *Seafarer* and the *Balsa 37* watchkeepers used the VHF radio to arrange to an unorthodox, last-minute starboard-to-starboard meeting just before colliding. We don't know if collision could have been avoided by simply sticking to the original and more conventional port-to-port agreement, but it raises an important question concerning the role of verbal communication in collision avoidance. Should verbal communication be avoided? This view has been endorsed by some regulators, and in Britain the following notice has been issued:

> There have been a significant number of collisions where investigation has found that at some stage before impact one or both parties were using VHF radio in an attempt to avoid collision. The use of VHF radio in these circumstances is not always helpful and may even prove dangerous.... Valuable time can be wasted while mariners try to make contact on VHF radio instead of complying with the Collision Regulations.... Any attempt to use VHF to agree on the manner of passing is fraught with the danger of misunderstanding.

This is strong counsel intended to temper the reflex to turn every encounter into a conversation at the expense of following the Rules. It should also be appreciated that in high-traffic areas, every radio call is a potential distraction for everyone in earshot. Keeping radio chatter to a minimum is good sea manners. Additionally, in multivessel encounters there is always the risk that you are not talking to the vessel you think you are talking to.

The counterargument is also strong: If you can confirm the intentions of another vessel and eliminate ambiguity with a quick radio call, then you should. Another advantage of the proactive use of radio is that a call can serve as a scouting report in that you can learn a lot from the response, if you get one. Does the other person have a command of English? Is he or she aware of your presence? Does the person sound alert? Does the response reflect a grasp of the situation and the Rules? Is the watchkeeper even listening to the radio? The answers to these questions can inform the level of caution you apply to the encounter.

Each of the arguments above concerning radio use has its merits, but the debate illustrates the manner in which VHF radio, a cornerstone of marine communications, can be a double-edged sword.

Effective Communication Techniques

The main point of shipboard communication is to get essential information where it is needed in a timely fashion. Decision makers can only make good, informed decisions when they have good information. This simple fact puts communication at the heart of BRM. Both senders and receivers have significant responsibilities in communicating, whether upholding them comes naturally or not. Even though power and decision making are vested in the upper reaches of the chain of command, the alternating quality of a two-way exchange means that communication itself it is not hierarchical at all. If a deckhand has just seen something important that no one else has seen, then at that moment he or she is the most important person on the ship. If captains and officers see themselves only as senders and people farther down the chain see themselves only as receivers, sooner or later critical information will not get where it needs to go, and faulty decisions may result. As it turns out, the people with the most responsibility (and the most to lose) also have the greatest incentive to cultivate effective two-way communication. As leaders, they are also in the best position to do it.

Successful communication is often achieved through what appears to be common sense but, in fact, is just a set of good habits learned from example and experience. Think back to your first radio call. Most likely it was awkward, halting, maybe even embarrassing. But you got better. But what exactly improved? Since effective two-way communication is so central to a well-run ship, let's examine what goes into it. What should good communicators strive for?

The Loop

A *loop* is often used to describe a successful communication: a message goes out, it is received, and the receiver closes the loop with an acknowledgment. If you are not in the loop, well, you're not in the loop. While this image is simplistic, it illustrates many of the fundamental practices that go into good communication, including the following.

Build the loop: Establish who is in it, what is expected of them, and the means of communication to be used. With non-face-to-face spoken communication, establish an open line of communication by identifying who you are and where you are, and find out who you are talking to and where that person is. This helps you see things from the other person's viewpoint.

Maintain the loop: Competing demands are a normal part of vessel operations. In a bridge or deck team setting, look at what other people are doing and where their attention is. It does no good to issue an order, ask a question, or report traffic if the receiver's attention is elsewhere. Also, don't muddle the message or distract others with unrelated information.

Get in the loop: If you have information or a concern that is relevant to the situation, don't keep it to yourself. Get it in the loop.

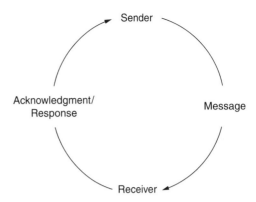

In a shipboard communication loop, the Sender is not always the captain/boss, and the Receiver is not always the crew/subordinate.

Stay in the loop: Pay attention and minimize distractions. If communication is the highest priority at the moment, then prioritize effort accordingly.

Close the loop: Don't leave people hanging. Close the loop by acknowledging or confirming the message. Without this, things can only move forward on the basis of assumptions. An unanswered message sows uncertainty.

The loop can be strengthened by applying the following tips:

Explain your intentions: Create the necessary mental model. Don't assume everyone knows you plan to overtake if you didn't make it clear to everyone.

Say it twice (or more): There may be interference or unfamiliar details that the receiver can't grasp the first time around. Saying it twice can get the message through better than if you have to resort to a series of clarifying exchanges.

Clarify and Confirm: If you do not understand the message, either the words themselves or their meaning, request clarification. And ask for confirmation if it is not forthcoming: "Do you copy?" or "Do you understand?"

Follow up: After a message has been sent and confirmed, watch to see if the actions of the receiver are consistent with the message. This could mean glancing to see if the helmsman is responding correctly to a course change or if another vessel is complying with what was agreed upon. Following up is a last line of defense against misunderstandings, and, occasionally, it is how you get your first clue to an error chain.

Last, here are two communication fundamentals that trump all of above:

Listen: Above all, listen. No one learns anything when they're talking.

Talk the talk: Officers are leaders and will be emulated. Model the communication approach that you expect and require. If you expect a crew member to repeat a command, then you owe them an acknowledgement when they do it. That's part of closing the loop.

Language and Terminology

Two hundred years ago, the side of a vessel opposite the starboard side was referred to as the *larboard* side. After what must have been an unspeakable number of misunderstandings, someone decided to change larboard to *port*. The name stuck, and we are thankful to this day. Effective communication requires using words and phrases that are readily understood, not ambiguous. The nautical world contains a wealth of specific language for this very reason, and professional mariners should use it.

Even so, a common language can be surprisingly elusive in an industry that goes to great lengths to standardize. "Standard" maritime English may not work in places where local vernacular hold sway. For example, in much of the United States the phrase "See you on one" is accepted code for a conventional port-to-port passing agreement, but that phrase is not universal. While most mariners could decipher the intent, the fact remains that saying "Meet you port to port" is more to the point.

As noted, ships travel, so even if you are local, the person you are talking to may not be. With this in mind, pick your language carefully because a communication failure cuts all ways. Pointing fingers after an incident is small compensation for damaged careers, damaged boats, and injured people. Try the following recommendations to help you communicate with effective language and terminology.

Be professionally accurate and precise (within reason): Accurate, precise language leaves less room for misunderstanding. In some cases it may be better to say "I am the northbound vessel" even though you are steering 018 degrees. On the other hand, announcing that "I am the inbound fishing vessel" does little good if there is more than one inbound fishing vessel within VHF range. Context matters to your choice of language and terms.

Avoid slang and vernacular (within reason): Some will understand it, but others may not. If there is a chance it will impede communication, don't use it. That said, do what it takes to be understood. Don't let dictionary definitions get in the way of getting your point across. The true test of any communication system is whether or not it works.

Be consistent: A bow spring line, an after-leading spring line, a forward spring line, and a number 2 line can all mean the same thing. The accepted term at any given moment depends on what boat you are on and who the captain is. The same goes for right rudder versus starboard rudder. While you always know what *you* mean, consistency helps others to know what you mean too.

Keep It Simple, Sailor: Use straightforward, familiar phrases and words, such as the following:

- Yes
- No
- Roger
- Okay
- Aye
- Stand by
- Understood
- Leave it to port (or starboard)
- I have the conn/Do you have the conn?
- I concur/Do you concur?

Communication Protocols

As with all professional language, established communication protocols and points of etiquette help people of different backgrounds quickly establish mutual understanding without reinventing the wheel or losing precious time. An effective communication protocol anticipates the potential for misunderstanding, and therefore is structured to prevent as many as possible. To be on the safe side, it also includes ways to discover misunderstandings and undo them. The phonetic alphabet, in which each letter is associated with a specific word (A = Alfa, B = Bravo, C = Charlie, etc.), is a good example. Established communication protocols have a way of mutating according to local convention, but if you don't know them at all, you can't use them when needed. Here are some examples of standardized communication protocols pertaining to watchkeeping.

Point system for identifying relative positions of objects: Saying "Vessel two points on the starboard bow" is clearly superior to saying "There's a boat over there." When contacting another vessel, always describe yourself from their perspective: "I am the eastbound vessel on *your* port beam." The terms *quarter*, *bow*, *broad*, and *fine* are all clarifiers for this type of communication.

Use three-digit notation for courses and bearings: "045°" spoken as "zero-four-five degrees."

Position: Use a range and bearing from a known location: "Calling the southbound vessel approximately one mile east of Point Lookout." If you must rely upon latitude and longitude, give it in degrees and minutes, with latitude first, followed by longitude. A lat/long position is awkward over the radio but under some circumstances there may be no way around it. Be prepared to repeat the position.

Charted place names: Use place names as they appear on the chart. Unofficial place names can be useful, especially when there is no official place name. But when a local place name is unfamiliar to some of the people involved, there is clear potential for confusion.

Time: Twenty-four-hour time is standard for most round the clock operations (1300 not 1 p.m.), but that system may not prevail in all situations.

Helm commands: There are countless variations and preferences around the world, but internationally recognized standards do exist and should be used whenever appropriate.

Official maritime language: This is English. You should use it, and expect it to be used, in multilingual situations.

Style and Structure

Effective communicators get points not only for content, but how they deliver it, in other words, style. Style may be a bigger factor when face to face than it is in less direct forms of communication, but the delivery always has ramifications for the success of the message. Here are some things to consider with regard to the structure of a message and the manner of delivery.

Control the volume: On a ship, people should never have to strain to hear you. If people are asking you to repeat yourself when there is no interference, then there is a problem with the delivery. Speak clearly—enunciate, and don't mumble.

Control the rate of delivery: In general, there is a greater chance for misunderstandings when people speak too quickly. Deliver information at a rate that the receiver can handle. This is especially important in a crisis situation, when there is a tendency to speed up.

Control the "bite" size: If your message contains multiple steps or points, deliver them in small bites, and get confirmation at each stage. There is no point in delivering a lengthy monologue if the receiver is lost after the first few sentences.

Control the sequence: When your message has multiple parts, transmit them in a sequence that is consistent with the situation. If getting the anchor ready is the highest priority of three tasks, don't tell a crew member to first wake the next watch, then check the engine room, and finally get the anchor ready. Using a logical sequence requires that you compose your thoughts before you articulate them. Be aware of the possibility that what appears logical to you may not seem so to someone else. You may have to rephrase or clarify.

Combine modes of communication: Since visual cues are an important component of face-to-face communication, use them to your advantage when possible: speak, point, imitate, sketch it out, and use intonation and gestures. Make the most of these resources.

Make eye contact: When possible, make eye contact with people. It gets their attention, which is what you want.

Make space: In addition to listening, make space for others to speak. Some people won't fight to be heard. If you leave some space, you might hear something important.

Keep it clean: Avoid profane, derogatory, or offensive language. Such language is a distraction that carries other "messages" that rarely benefit the exchange.

Keep it lean: Although the crew on the *Seafarer*, the *Balsa 37*, and the *Bouchard* didn't communicate enough, we have all heard radio exchanges in which people said way too much. Be disciplined, stay on topic, avoid rambling, and avoid pointless verbal crutches that ultimately add nothing and may detract.

Be courteous: An order is not a request, and it should not sound like one. The opposite is also true, however. There are few professional communiqués that cannot be delivered with courtesy while also conveying authority and command. Striking this balance does not always come naturally, yet it is an essential measure of competence for a leader. Some cultures and some individuals place such emphasis on politeness that it can get in the way of getting the point across. This can be counterproductive.

Wear the other person's "shoes": If you want to succeed as a communicator, you must try to put yourself in the other person's shoes. This is easier said than done, but it is a vital part of successful communication.

Communication, like most endeavors, benefits from thinking ahead and practice. In emergencies communication is more important than ever, but the impediments may also be greater. When emotions and stress run high, it takes great composure to stick to the good habits that come easily on a regular day. At such times, we must exert discipline in the tone, style, and structure of our communications. Even when you don't *feel* calm, you can at least try to *sound* calm. This, too, is part of leadership.

The cost of misunderstandings can be high. Every communication, no matter how straightforward, no matter how clear, plain, obvious, routine, or familiar, carries with it some potential for failure. As long as mariners keep this fact fresh in their minds, they are more likely to make the extra effort to ensure that the intended messages get through.

Between the Lines of Communication

Despite our best efforts and the use of the communication techniques we've discussed, we all know that communication breakdowns happen. Investigators in the *Herald of Free Enterprise* case closely questioned the second mate, who was loading officer on that fateful day, and the first mate, who took over the loading, to ascertain how that interaction might have impacted the failure to close the bow doors. We know that there was some tension between the two officers, and that they did not meet face to face to transfer the loading officer responsibilities. The second mate was not formally relieved—the first mate simply started issuing orders over his radio. "I knew the job had been taken away from me," the second mate explained. "He took over as loading officer so I assumed he took the responsibilities that go with that job." This explanation sounds tinged with indignation; yet why did the first mate take over in this manner? If it was part of a simmering personality conflict it only proves that they were human. When communication between people who are responsible for the lives of others suffers as a result, then we have a problem.

Human beings broadcast subliminal messages on a wide range of "unofficial" frequencies. You don't need a degree in communications to know that a person who slams a door is probably angry. When someone nods his or her head, we generally take that to mean that the person understands or agrees with what is being said; yet most of us have been guilty of nodding our head when we did not understand or agree at all. Where hierarchy is paramount, people may nod their heads no matter what because to do otherwise is socially unacceptable for one or both parties. We learn early in life that communication is more than words. We should apply this knowledge professionally because it matters.

Situational Awareness

Virtually all the circumstances that degrade situational awareness also serve as barriers to effective communication. Distraction can pull a person out of the communication loop at a critical moment and interfere with sending or receiving clear, complete messages in time. Stress can do much the same, but it can also prompt people to internalize their thoughts and withdraw when they should be interacting. Stress can also prompt mariners to adopt a sharp tone under pressure that alienates allies and counterparts, and undermines the collaborative spirit. Fatigue has the capacity to bring about any of the above.

It is hard to treat the ordinary as significant. Often it is the very ordinariness of communication—in other words, complacency—that leads to breakdowns. We saw this in the Tampa Bay incident where, despite the experience of everyone involved, communication was pursued by half measures. Overreliance diminishes situational awareness when the equipment that transmits and receives vital information fails to work or mariners fail to use it properly. Transition undermines situational awareness by spreading cognitive resources too thinly or focusing them improperly. If situational awareness is low for any reason, communication breakdowns are more likely.

Bias

William Faulkner wrote, "The past is never dead. It's not even past." History, both personal experience and the hand-me-down variety, provides the lens through which we view the world. We are each faithful prisoners of our own special perspective. This is not always bad; different ships, different long splices. But when history, perspective, and bias get in the way of good shipboard communication, we have a dangerous situation. In Chapter 9 we discussed the ladder of inference, and how people have a tendency to swiftly climb from one assumption to the next, sometimes arriving at seriously flawed conclusions. Nowhere is the ladder of inference more applicable than in the arena of human interaction, where assumptions and bias can steal attention from where it belongs, disrupt lines of communication, and derail cooperation between people who need to work together. Here are a few well-known biases that aspiring mariners can expect to encounter, and which they will have to communicate around or through.

Professional bias: The common ground that mariners of all stripes share, and that sets them apart from the rest of the world, should produce an unbreakable bond of fellowship, right? Well, sometimes it does, but often professional prejudices can jeopardize lines of communication when the only thing that should matter is a safe passing agreement or clean change of the watch.

An example of unhelpful professional bias is the needless static based on generalizations that people in the towing, yachting, offshore supply, passenger-carrying, commercial fishing, sailing, ship-assist, piloting, deep-sea merchant marine, Navy, and Coast Guard sectors make about one another. Another divide runs between hawsepipers and academy grads. If a crew is large enough, it can accommodate departmental splits (deck versus engine.). The only thing capable of melding these underway antagonisms into a more universal contempt is the disdain normally reserved for regulators and possibly shoreside management. Whether these characterizations are the exception or the rule, or just a source of humor, if they get in the way of the message, or become the message, it can lead to a communication breakdown with consequences.

Hierarchy and "yes-manship": Hierarchy has a special place in maritime culture; yet some cultures, institutions, and individuals are more inclined than others to see it as absolute. Most would agree that hierarchy is useful for clarifying lines of responsibility and communication and for establishing a level of formality that is appropriate to serious work. But when blind devotion to hierarchy produces an atmosphere of "yes-manship," it can be counterproductive. This can occur when people unquestioningly agree with leadership or outwardly agree to avoid reprisal. A hierarchy that upholds an atmosphere of condescension, brownnosing, and reprisal can be an insurmountable barrier to productive communication. When people are unwilling to communicate honestly (which includes listening) in the professional arena, good resource management and a culture of collaboration are unattainable.

On the tug *Seafarer* entering Tampa Bay, the off-going mate's situational awareness was apparently higher than that of the oncoming captain, but it didn't seem to bear upon the captain's decision making. In seafaring culture the captain pretty much does what the captain wants because he's the captain. We can see in retrospect however, that the decision to overtake called for skepticism, and it was not forthcoming.

We don't know exactly why the chief mate on the bridge of the *Balsa 37* became so irrelevant in the presence of the pilot, to the point of not even listening to the VHF radio. Had the pilot been intentionally or unintentionally dismissive of the flag-of-convenience chief mate? Did the chief mate's background or personality predispose him to be too deferential to the pilot even in the face of collision, or was he just tired? We will never have complete answers, but if the goal of communication, and by extension BRM, is to push good information to where it is needed, then rank should not get in the way of that.

Cultural, social basis, and personality: Obviously, significant cultural and social biases can exist around gender, race, ethnicity, place of origin, and other factors. Depending on where a person works and what their background is, these things can be a big deal or not. When social or cultural intolerance handicaps effective communication and hinders professional cooperation on the water, then it calls into question whether a person is actually qualified to be an officer.

Sometimes the first clue we get that we are dealing with the "other" is the sound of their voice over the radio, a communication mode that we already know is second best. Just a few words can trigger assumptions that distract from the message. If this interferes with the message, the resulting communication breakdown can be just as significant as a faulty sound-powered phone, only much harder to repair.

Outside the issues mentioned above, sometimes you just don't like someone. Intolerance based on nothing more than personality can be every bit as detrimental to effective communication. When these situations arise, and they inevitably do, we cannot let them derail good communication and doing the job right.

What all these fault lines share is a measure of mistrust. When a lack of trust hangs heavy on the lines of communication before the first word has been uttered, then failure is a strong possibility. People will see what they want to see in others, and that is their right—up to the point where professional conduct and safety suffer.

*****

A good working relationship is both a product and a reflection of good communication. When people are familiar with a task, as well as a person's accent, phrasing, idiosyncrasies, strengths, and weaknesses, they are better able to compensate when communication lapses do occur: they know what you *meant* to say. Assumptions can play a useful role in simplifying communication but they can also take us too far, and in the wrong direction. Moreover, mariners don't always have the advantage of established working relationships. Part of successful communication on the water is the ability to initiate an effective working relationship on short notice. This is what we must do every time we contact an unfamiliar vessel or welcome a new crew member aboard.

Effective communication sometimes comes easily, but it is not by chance. The better we understand what goes into it the more likely we are to repeat what works and avoid what doesn't. Making appropriate choices according to the overt lines of communication, while paying attention to the subtext between the lines, can help minimize communication breakdowns.

Chapter Twelve

Teams and Teamwork

SHIPBOARD TEAMS DO NOT EXIST because playing on a team is more fun than playing alone. Shipboard teams exist because they can enhance the situational awareness of the team leader by providing more perspectives, and more capacity for problem solving. Shipboard teams limit single-point failures by catching the errors of individual team members; if one human component fails, the entire system may not. Indeed in a 5-year period ending in 2005, the Marine Accident Investigation Branch (MAIB) in Britain found that nearly 60% of all collisions and groundings could be attributed to single-handed watchkeeping. A team can supply expertise that an individual may not have. A team makes it possible to distribute workloads when there is more to do than one person can manage alone. A well-conceived team can relieve pressure on the team leader and act as a bulwark against poor decision making. Implicit in shipboard teamwork is the concept of specialization coordinated toward a common goal, something more like an orchestra than a cavalry charge. Indeed, individuals think and act differently when they know that they have a specific role within a team, as opposed to when they act independently or have no defined role at all. By allocating responsibilities within a group of people, the sum can be greater than the parts, or so the thinking goes. The ability to specialize largely depends on the resources available. In general, vessels with smaller crews can afford less specialization than vessels with larger crews.

Teams can overcome many of the factors that contribute to human error, although sometimes they can create their own problems. Like the people in them, teams are imperfect. Teams have more moving parts, which means more opportunities for mistakes and misunderstandings. Working with and through others creates a communication burden and requires team members to think and act with more lead time. Many competent boat handlers accustomed to personally handling the controls can become flustered trying to accomplish the same task through verbal commands to a helmsman. The knowledge is still there, but the process is different. A good team can do more than any individual, but a flawed team may be able to do less. Teamwork and leadership are indivisible concepts, and a team without leadership is chaos.

Sometimes teams emerge naturally, making successful teamwork appear inevitable. No one wants to participate in failure, and the benefits of a team approach are often obvious. But as much as possible, shipboard teams should be intentional rather than spontaneous. The best teams contain people who are conscious of the team dynamic and strive to work within it. Here we are most interested in bridge teams, but the same collaborative spirit can, and often must, extend beyond the wheelhouse and even beyond the ship. You don't have to be standing next to someone to experience teamwork or its absence. Teamwork is an important piece of BRM, even for vessels operating with small, tight-knit crews where specialization is more limited. Fortunately, much of shipboard life promotes teamwork: the chain of command, defined roles, a sense of community, and the opportunity to hone performance through repetition. But a crew, by virtue of being a crew, does not necessarily make a team.

The *Empress of the North*, 2007

The *Empress of the North* was built in 2003 along the lines of an old-time paddle steamer, but fitted out to meet the needs of contemporary passenger carrying. On May 12, 2007, the vessel arrived at Juneau, Alaska, to begin the summer season. On that same day a new third engineer and a new third mate joined the ship, both having graduated from a maritime academy two weeks earlier. Recognizing that the new officers lacked experience outside the training environment, arrangements were made to bring them up to speed. The first engineer rearranged his schedule to stand watch with the third engineer, and the company built

The *Empress of the North* underway in Auke Bay, Alaska. (National Transportation Safety Board)

in a seven-day orientation period for the third mate who would shadow the more experienced third mate to learn the ship, equipment, routines, and the waters before assuming the responsibilities of a watch officer. Putting the new third mate in charge of a watch in confined waters on an unfamiliar vessel, perhaps at night, without preparation or supervision was considered imprudent. While no amount of preparation can eliminate the steep learning curve that accompanies those first few watches, at least there was a plan to manage the new third mate's transition into the real world.

The ship sailed for Skagway on the evening of May 12 with 206 passengers aboard. Shortly after midnight on May 13 the senior third mate became ill while standing the 0000 to 0400 watch. The chief mate took over, and then stood his own watch until 0800. Since the new third mate did not stand watch at all that first night, it wasn't until the morning of May 13 that he learned about the senior third mate's illness. At noon the new third mate took the 1200 to 1600 watch while dockside at Skagway. During this period the captain informed him that due to the other third mate's illness, he would need to cover the 0000 to 0400 watch, which was an underway watch.

After completing the dockside watch the new third mate assisted the chief mate with pre-underway equipment checks, went to dinner, and then retired to his cabin. He arrived on the bridge at about 2320. Despite the change of plan, at no point prior to coming on watch did anyone brief him on the route, orient him to the bridge equipment, or provide any specific guidance or expectations in the wake of his "battle-field promotion." Instead, the captain rearranged the watch assignments so that the most experienced able-bodied seaman (AB) would stand watch with the new third mate in addition to the regular AB and the ordinary seaman (OS). According to the captain, the AB was working toward a license and was knowledgeable about the bridge equipment. He was thirty-six years old and had been with the company about seventeen months, initially as an OS. He had traveled the route once before, though not as an AB with helm duties, and not recently. The captain described the AB as "unintimidating," a quality that would presumably facilitate a productive partnership with the new third mate and compensate for the new third mate's inexperience. The fact that local pilotage requirements called for watch officers to have at least four round-trips in those waters, one of which had to be at night if the ship was operated at night, was not addressed.

When the new third mate arrived on the bridge, the second mate proceeded to hand over the watch for what would be the first underway watch on the third mate's new license. During the 20-minute overlap, no less than seven people visited the pilothouse, two of whom had no watchstanding duties at all, despite the fact that operating procedures and federal regulations prohibited crew who were not engaged in watchstanding from being there. Both of these crew members conversed with the two mates on unrelated topics during the handover. The AB, who had been tapped for his experience, did not arrive till the handover was complete and the second mate was departing.

This view of the bridge of the *Empress of the North* clearly shows the Nobeltec monitor with the ECDIS monitor in back of it. (National Transportation Safety Board)

MAJESTIC AMERICA LINE -----FLEET INSTRUCTIONS Empress of the North			
Date Issued: 06/03/04	Date Revised: 06/03/04	Approved By: Randy Burns	

STANDING ORDERS   M/V EMPRESS of the NORTH
☞ **OBSERVE THE RULES OF THE ROAD** ☜

1. The orders below must be read by each watch officer before taking their first watch on the bridge. He/she must sign on the appropriate line indicating that he/she understands these orders.
2. When alone on the bridge you should keep in mind that the time to take action for the vessel's safety is while there is still time.
3. An officer should be on the bridge at all times when the vessel is underway.
4. Before relieving the watch, the relieving officer will sign the night orders, acquaint themselves with the vessel's position, course and speed, weather conditions, and any contacts visual or by radar and obtain any pertinent information the officer to be relieved may have to pass on.
5. As watch officer of this vessel you are, when on duty, expected to keep a good watch and see to it that your lookout does the same.
6. When visibility becomes poor or if you anticipate that visibility may become poor because of fog, mist, rain, snow, or any other reason, call me. In the meantime post lookouts, start Fog Signals, reduce speed to safe speed. If necessary, post one lookout on the bow with a radio for deck to bridge communication.
7. Be sure your lookout is thoroughly familiar with his/her duties, that they keep alert and that they listen as well as look out for other vessels and possible hazards. He/she is not to be assigned other duties. When your personnel are on lookout, they should be properly dressed for the weather. If the weather is inclement, keep the lookout on the bridge. A lookout protected from the weather will keep a better watch.
8. **CALL THE CAPTAIN ANY TIME WHEN IN DOUBT**, but do so in ample time—better too soon than too late. Make sure the vessel is safe. Call me if the weather starts to make up or you think it might be necessary to change course or slow down. Do not allow the vessel to pound the seas.
9. Give passing vessels a good berth in ample time. DO NOT try to bluff the other vessel out of their right of way. Let the other vessel know in plenty of time what you intend to do.
10. Whenever underway, the radar must be turned on. Do not wait till fog or any other cause to shut around you. If using the short range, be sure to switch to the long range and intermediate ranges periodically.
11. Whenever underway, take a fix at a minimum of once every hour (radar, gps, visual bearing) and log it in the underway deck logbook. Always verify a fix taken with one type of gear, with information from another source and confirm sounding on the chart with the ship's depth sounder. If there is discrepancy with regards to the ship's position or the observed depth, call me.
12. Check the course on the chart every time there is a change of course on your watch. Call the Captain immediately if there is an error or the course will lead the vessel into danger.
13. This vessel is to be put into hand steering at all times a Pilot has the con, when maneuvering through a bridge and its supports, approaching or maneuvering in locks, when within 0.5 NM of another vessel in any situation, and when you, as the officer on watch, think it necessary.
14. Call the Captain if you experience an unexpected power, steering, or other equipment failure, or if the ship begins to drag anchor, or you see or hear any type of distress signal.
15. Give engineers a minimum of 15 minutes notice prior to maneuvering alongside if the bowthruster is required.
16. Always check steering and engine controls prior to getting underway and maintain an active watch on VHF radio channels 16, 13, and any other locally applied traffic channels (as in 14 for locks).
17. Call the Captain when anchoring at night or in limited visibility (i.e., fog, rain, snow).
18. If it becomes necessary to take the vessel outside the buoyed or charted channel due to ship traffic, call the Captain immediately. Other than to let large ship traffic pass, stay in the channel.
19. Keep all log books up to date while on watch. Entries in the radio log should be made by the operator in regards to a listening watch.
Log all significant communications made on either VHF or SSB radios. This includes VTS check-in and out, Distress calls and all communications with any government agency.
20. Any significant change to the ship's itinerary (ie. Change in the ship's destination or a significant change to the ship's departure or arrival time) must be approved by the Captain prior to any implementation of said plans.
21. A proper anchor watch is maintained by taking hourly observations of ranges prominent to landmarks, water depth, and wind speed. Periodic observations should be taken in excess of hourly notations.
22. The officer of the watch is the Master's representative and as that representative, the safe navigation of the ship is his/her primary responsibility. He/she should at all times comply with applicable regulations for preventing a collision at sea. It is essential that the officer of the watch appreciate that the efficient performance of their duties is necessary in the interest of safety and life and property at sea and the prevention of pollution of the marine environment.
23. Please sign below indicating that you understand these standing orders, which will be supplemented by night orders each evening when the vessel is underway or at anchor when required. In signing you are also acknowledging that you have read and will comply with these guidelines. A copy of these Standing Orders is to be kept in the front cover of the logbook for reference.

MASTER	DATE
CHIEF MATE	DATE
SECOND MATE	DATE
THIRD MATE	DATE

The standing orders for the *Empress of the North* at the time of the Rocky Island grounding. (National Transportation Safety Board)

The pilothouse was well equipped, and, in general, the waters were deep and free of hazards. However, the route included one critical choice at Rocky Island. You could go either north of Rocky Island and transit the half-mile-wide slot between the island and shoal water off Point Couverden, or take the slightly longer but more open route south of Rocky Island. At 12 knots, the northern route shaved 6 minutes from the transit time. The track lines appearing on the charts were ambiguous as to which route to take. One electronic chart display showed the planned route running north of the island, through the slot between Rocky Island and Point Couverden. The ECDIS showed the planned route transiting the open water south of Rocky Island. The paper chart showed both. The passage plan included no navigational guidance or parameters, and was oriented toward passenger activities. There were no night orders and no "Call Me Here" points. If the new third mate had consulted the *Coast Pilot* he would have found the following helpful information: "As currents are erratic in the channel between Rocky Island and Point Couverden, the slight saving in distance . . . does not warrant its use." The second mate recounted his discussion with the new third mate of the route this way: "I told him that he needed to go south of the island if he didn't feel comfortable, or he could go right down through the channel there. He didn't tell me what he was going to do either way. He just nodded his head."

Ambiguity on a ship is never helpful, but when it is ensconced in the passage plan it can be especially disorienting to people with unfamiliar duties. As it turned out, this bridge team would attempt to go both ways around Rocky Island.

The standing orders required all watch officers to have "a thorough understanding of bridge equipment and electronics and their use," but there had not been time for this; therefore the AB, who was functioning as the helmsman, began orienting the new third mate to the bridge equipment. The idiosyncrasies of the alarm panels, Z-drives, electronic charts, radars, and VHFs were all unfamiliar: "Almost everything but the paper chart and triangles," the third mate told investigators. The new third mate was nominally in charge, but in his mind it was not so simple: "My understanding was he [the AB/helmsman] would be in control of the boat, and I would be there because I have a license. . . . I would assist in what he needed me to do, and learn."

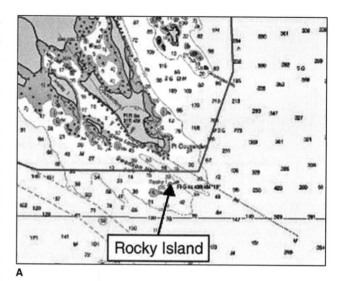

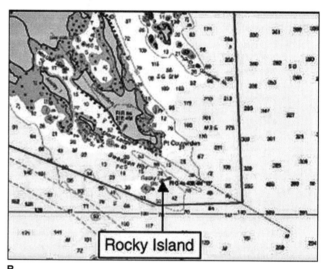

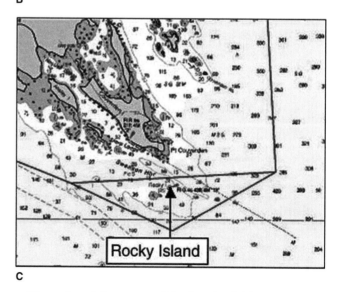

Charted track lines around Rocky Island. The Nobeltec display indicated a northerly route (A), the ECDIS indicated a southerly route (B), and the paper chart showed both (C). (National Transportation Safety Board)

> **EMPRESS OF THE NORTH**
> *2007 Alaska Itinerary*
> *Day 3-Monday*
> *Glacier Bay/Pt Adolphus*
>
> | Time | Location/Comments |
> |------|-------------------|
> | 0330 | *Security call for rounding Rocky Is, west bound Pt Couverden* |
> | 0545 | Bartlett Cove. Pick up Rangers |
> | 0600–? | Glacier Bay |
> | 1730 | *Security call for east bound Icy Straits* Underway, 144 NM to Sitka, 73 NM to Morris Reef, 112 to Sergius Narrows *GPS Route: AK GBNP 2 Peril* |
> | 1900 | Pt Adolphus   Whale Watching |
> | 2030 | Underway for Sitka *Max speed to Morris Reef then adjust for current at* Sergius Narrows *Security call for rounding Pt Augusta, east bound* |
> | 2200 | The Sister |
> | 2300 | Point Augusta |

The passage plan for the *Empress of the North* on day 3 of a regular trip emphasized passenger activities, not navigation. (National Transportation Safety Board)

At about 0100, the other two crew members assigned to the watch left the bridge to make a routine security round of the ship. Their departure meant that the AB/helmsman, who was engaged in tutoring the new third mate, also now had to function as lookout, a violation of the STCW code and the COLREGS. Moreover, his unspecified advisory capacity began to expand into functions that are normally the purview of a licensed officer, including making a passing agreement with another vessel and taking radar ranges and bearings.

At 0126, with music playing in the background, the third mate and the AB/helmsman spotted the flashing green light that marked Rocky Island. At 0127, the AB/helmsman began to make the 80-degree turn to starboard that would bring the ship north of Rocky Island and through the slot. He did this entirely on his own initiative, without orders, applying rudder as he thought appropriate. The third mate said later, "I relied upon his expertise as to when and how much to turn." Soon the AB/helmsman articulated a concern that the vessel wasn't turning fast enough, so he increased the rate of turn, again with no input from the new third mate. Then, again without instructions or announcing his intentions, the AB/helmsman altered course to port, signaling a decision to go south around the island. As the vessel swung first one way and then the other, the new third mate lost track of the flashing light on the island, and there was no dedicated lookout to assist in tracking it. Suddenly, the third mate saw rocks to starboard. At this point he attempted to take charge: "Hard left!" A few seconds later the ship grounded on Rocky Island, literally the only isolated danger for miles around, with the radar still on the six-mile range scale.

The captain arrived on the bridge within seconds and began to direct the emergency response. Though five watertight compartments were damaged, flooding was limited to tanks and void spaces and did not progress through the boat. The vessel adopted an 8-degree list but remained afloat. Within an hour of issuing a Mayday call, the first of several good Samaritan vessels arrived at the scene and began removing passengers. Life rafts were launched, though the captain considered these a last resort due to the large number of elderly passengers, some of whom were in wheelchairs. It was nearly three hours before a Coast Guard cutter was able to reach the scene, and, since many of the responding vessels were small, it took almost four hours to complete the evacuation of passengers and nonessential crew. The fact that everyone survived this episode and that the vessel did not sink should not obscure the fact that a large number of people came very close to ending up in icy waters, in darkness, far from help. Under such a scenario, many people might have lost their lives.

Failure of Teamwork

From the moment the captain decided to solve his manning problem by putting the new third mate on watch, an error chain was set in motion. From a BRM standpoint, almost nothing was right. The passage plan lacked detail. Procedures and regulations were violated on a large scale with regard to lookout, pilotage, who was allowed on the bridge, and familiarization with the bridge equipment and emergency procedures. There were distractions on the bridge during the watch transfer, the stress of a new job, and a situation defined by transitions. There was a lack of direction regarding the route and the captain's approach to risk assessment showed signs of complacency. For all this, it was the absence of a functioning team when it counted most that sealed the ship's fate.

It may seem from this presentation of events that an accident was inevitable, but that's not true. In the real world there are far more close calls than consequences. This fact shapes behavior toward risk, and undoubtedly influenced the captain's decision to put the new third mate on watch that night. Rest assured, if every young officer who was in a little bit over his head put a ship aground, a great many things would be done differently. In this case, an uneventful watch would have reinforced the illusion of the new third mate's readiness to take charge, calling into question the value of having any kind of overlap at all. While there were serious consequences of this incident, nobody died; therefore it can be said that the lessons were purchased at a discount. However, repairs to the *Empress of the North* came to nearly $5 million, and the company lost almost two months of revenue. The captain's license was suspended, and both he and the new third mate were fired. The turmoil the incident caused led to a total rewrite of safety and operational policies, followed by bankruptcy.

The third mate did not conduct himself on the bridge as an officer should, even a new one. By ceding de facto control to the AB, he betrayed the responsibilities of his license. Whatever he was told his responsibilities would be, and whatever he conceived them to be, he never grasped that he was in charge. An AB cannot take responsibility for a navigational watch, no matter how knowledgeable he or she may be. In accepting the unorthodox watch arrangement in place of the original plan to bring him up to speed gradually, the new third mate undercut himself. Trial by fire is an old tradition of the sea, often recounted with gusto years later. But trial by fire carries the risk of being burned, and when this happens there is no joy in the memory.

Clearly, the captain put the third mate in a difficult position. How does a new mate say "no" to an experienced captain while teetering on the threshold of his very first job? And what would the ramifications be if he did? A principled stance might jeopardize his job. As it was, the real world found him six days ahead of schedule. The real world has a way of doing that. The ship needed him, and he wanted to be a team player.

Perhaps more important than itemizing the specific watchkeeping failures of a young, untested mate is the need to appreciate that his situation, while difficult, was not beyond him. His mistake, if we can call it that, was in not seeing the danger in the captain's plans, or in not having the guts to question the plan. To an extent he can be forgiven for this; the captain didn't see the danger either. The captain had a problem and the shiny new mate standing there with ink drying on his license looked a lot like a solution. This situation is likely to arise again someday, somewhere, and this incident provides an important perspective on how such a situation can end. When experience (or credentials) do not match the job description, stepping up for the good of the team may not be best for the team after all, and certainly not for the individual if things go wrong, no matter how flattering the opportunity may appear and no matter how it is presented. After all, this was a holiday cruise, not an emergency. There were other options, and we will look at those in the next chapter.

Types of Watchkeeping Teams

There are various ways of describing watchkeeping teams. Here we'll call them standing teams, the team-of-one, and mission teams. Each has its place, and particular advantages and disadvantages.

Standing Teams

Standing teams are the normal watch arrangement, whatever it may be. These are established teams that work together on a regular basis. The participants know one another's habits and tendencies, as well as the style and expectations of the leader (the watch officer or captain). There are less likely to be communication breakdowns because the context and the messages themselves are familiar, even anticipated. Overfamiliarity can lead to complacency, but on balance that risk is usually offset by the benefits of shared experience. When the members of a team know how things are supposed to go, they are more likely to detect something out of the ordinary.

Standing teams don't stand forever, of course. Crews change and watches are shuffled to meet the needs of the ship. Therefore, even standing teams undergo transitional periods during which they seldom operate as smoothly as before. New teams call for heightened situational awareness on the part of everyone, sometimes to the point of keeping a close eye on the newest member(s) of the team, even when that person is the team leader.

The bridge team on the *Empress of the North* was theoretically a standing team, but the inexperience of the third mate and the lack of continuity as a team left the team especially exposed to the hazards associated with transition. Even more significantly, there was a lack of clarity regarding roles within the team, and this was telling in the way events unfolded.

The Team-of-One Revisited

In the world of small vessels the team-of-one, which we first discussed in the Introduction and in Chapter 3, is an important subset of the standing team. The team-of-one isn't really a team at all, since it consists of a solitary watch officer who may have the support of an unlicensed crew member, but often doesn't. In this situation, the officer must specialize in all categories, allocating energy and attention accordingly. There's not much to say about team dynamics within a team-of-one, but one characteristic is worth considering.

With a solo watchstander, team navigation is a rare event and the wheelhouse is configured accordingly. A benefit of this situation is that it teaches self-sufficiency and working within one's limits. The problem with self-sufficiency is that a person can get used to it, leading to difficulties when the need for teamwork arises. In a team you have to trust others to do things you might normally do yourself. Instead of working at the speed of thought, you have to explain what you want or need. Not only does coordinating with other people require forethought, it requires energy. This is one reason people sometimes clam up in high-pressure situations: their cognitive resources become maxed out simply staying on top of details, and there is nothing to spare for communicating with the team. While the skill and experience that go into the team-of-one are impressive, teamwork benefits from practice. Going it alone all the time can be a handicap when a team approach is needed. Talk of teams may seem unnecessary to those accustomed to the team-of-one, yet sooner or later, perhaps in an emergency, all mariners end up functioning as part of a team, and it quickly becomes apparent if they are out of practice.

Mission Teams

Mission teams are assembled to handle a specific task that is more demanding than the normal watch structure is designed to handle. The shift from a standing team to a mission team may be triggered by traffic, weather, watch conditions, standing orders, or pilot integration. Mission teams are associated with less forgiving situations, such as close-quarters navigation, maneuvering, restricted visibility, or any circumstance requiring more resources and coordination than normal. This could, of course, include an emergency response.

While mission teams offer advantages, they are not without vulnerabilities. Although a mission team may be structured according to a plan (watch condition, station bill, etc.), it still entails a departure from the routine—in other words, a transition. Mission teams require people to shift gears, adapt, and interact in ways that are less familiar. Also, mission teams tend to be more complex entities, requiring more coordination. This is especially true when incorporating an unknown quantity, such as a pilot, into the team. To some extent, mission teams require individuals to temporarily reinvent themselves as a different cog on a new wheel, and some will be better at this than others. To offset these challenges, members of a mission team should be especially alert to the actions of others, and more deliberate in their communication than usual.

The Master-Pilot Exchange

A pilot is a resource, and a bridge team that includes a pilot constitutes a special sort of mission team. Pilot integration entails a unique transition that presents challenges for situational awareness, teamwork, and BRM in general. Many smaller vessels make use of pilots infrequently but when they do they are exposed to the same kinds of challenges that face larger vessels that never come or go without one. Mariners unaccustomed to integrating pilots may actually be more exposed to a team breakdown when they do than are those who are more practiced.

The *master-pilot exchange* (MPX) is a mechanism for meeting the challenges of pilot integration. On the face of it, the purpose of an MPX is to transfer factual information concerning the vessel and the transit as swiftly and accurately as possible in order to create a working mental model of how the transit will be accomplished. Indeed, this transfer of information is essential, but there is more to it. Inevitably, the MPX entails a certain amount of mutual sizing up as the pilot and the crew try to get a sense of who and what they are dealing with. The MPX, therefore, is a trust-building exercise between new, if temporary, business partners. The verbal exchange enables people to get a bead on manners of speech and proficiency with English and to establish standard terminology that will be employed in directing the vessel. Not only is the information exchange a vehicle for learning the newcomer's mannerisms and professional approach, but it can also give clues regarding the pilot's perception of risk, while transmitting the same things about your own operation. All of this is helpful to the partnership.

Obviously, an MPX that excludes important members of the bridge team will hinder the development of a more complete shared mental model. The MPX is not an isolated event, but rather an ongoing dialogue. It is the dialogue, not the initial exchange, that sustains the shared mental model and keeps the bridge team aligned.

The type of information addressed in the MPX varies with the vessel and operation, and there are a number of mechanisms for organizing this information. The Appendix details many of the points customarily considered in the MPX. Regardless of the specifics, the MPX should always extend to the following points:

Who's who: Communicate the names and titles of bridge team members and their roles and responsibilities.

Physical dimensions of the vessel: If a pilot card is available it must be accurate and readily understood. Blind spots (both radar and visual) and invisible distances (the area obscured by the bow) should also be noted.

Propulsion and maneuvering characteristics: A bridge poster, if available, can transmit basic maneuvering characteristics but it too must be accurate and readily understandable. If there is no bridge poster, then this information will have to be transmitted orally, taking care to avoid misunderstanding.

Bridge orientation: Even when pilots bring personal equipment (radio, electronic chart), they should be oriented to the bridge with regard to items related to the transit such as indicators for rudder angle, rpm, and rate of turn, as well as speed (water or ground) and heading (gyro, magnetic, electronic) readouts, charts, throttles, helm, ship's whistle, radios, etc.

Status and readiness: Detail the condition and readiness of the vessel, machinery, anchors, other equipment, and crew as they relate to the transit.

The passage plan: Review the plan, including anticipated tides, currents, visibility, restrictions, hazards, abort points, and contingency plans. The pilot may have new information that needs to be taken into account, not to mention a store of local knowledge.

Confirm the navigational situation: As with the change of the watch, pilot integration involves confirming the vessel's present position, course and speed, helm and engine settings, compass error, the traffic situation, and any other related information deemed pertinent.

Special concerns or considerations: Address any special concerns that might affect the transit.

In keeping with the idea of maintaining an ongoing dialogue, pilots should be expected to clarify their intentions, expectations, and concerns periodically throughout the transit. If something goes wrong, it is unlikely that it will be the sole responsibility of the pilot; therefore you need to know what is going on too. It is important to bear in mind that, like a watch handover, a master (or mate)-pilot conference can cover all the key points and still be ineffective. Pilots, while highly qualified, are fallible. This is another reason the bridge team must remain vigilant, despite the temptation to relax and let the pilot handle it. On rare occasions the pilot may cause the mate or the captain to lose trust in his or her abilities. Improper navigational decisions, lack of alertness, or incapacitation for any reason may make it necessary to relieve the pilot. This is an awkward and unusual situation, but it has occurred, and it *must* occur if the circumstances warrant. If it comes to this, it represents yet another transition that must be dealt with in time to reassert control over the vessel.

Attributes of a Successful Team

As is so often the case, we can make better decisions about what to do if we have an example of what not

Master-Pilot Discussion

Once the pilot is aboard, the discussion should include the following items, as appropriate.

Prior to Pilot Assuming Conn

Present heading/speed/position of ship

Last helm and engine order

Traffic in proximity

Show pilot location of critical gear (rudder indicator, rpm, indicator, speed log, whistle switch, VHF radio, etc.)

Prior to Commencing Transit (or as soon as practical)

Anticipated tides/currents

Draft and/or channel limitations

Maneuvering peculiarities or limitations of the ship including speed at which steerageway is lost

Maneuvering plan for channel transit including speed restrictions/limitations

Anticipated weather/visibility

Daylight restrictions

Status of navigation aids

New or existing navigation hazards

Other vessel traffic

Estimated transit time

Mooring plan or anchoring plan

Anchor readiness

Alternate maneuvering plans and emergency anchorages

Any problems with ship's gear

Compass error

Settings for VHF radio channels and radar ranges

Required flags and signals

If more than one pilot is embarked, which one is in charge?

Team Composition

The premise of a team approach in any endeavor is that a solitary person is not enough. So the first thing to figure out is how many people are needed. A rough approximation may be adequate in some cases, but where specific assignments (collision avoidance, external communications, position fixing, lookout, etc.) are involved, then the person in charge must make sure every base is covered. Under normal circumstances, an officer and an AB should have been able to conduct the *Empress of the North* along its route safely. However, because it was night, a designated lookout was also required. The watch started out with one, but as the ship approached the only part of the route that might be described as challenging, and where visual monitoring of the light on Rocky Island had real value, the designated lookout departed to conduct security rounds. The bridge team was depleted just as workload was increasing.

Numbers are only part of the equation. Qualifications, skill, and experience also enter in, and responsibilities should be distributed accordingly. Ideally, workload is spread evenly so that no person has too much or too little. The composition of the *Empress*'s bridge team did not reflect this. Clearly, the third mate's employer deemed him unready, thus the scheduled seven-day overlap. Additionally, the vessel was operating in waters that required either a licensed pilot or an officer with at least four round-trips. The new third mate had no trips. The AB/helmsman had one trip as an OS and didn't have the appropriate license. Additionally, the AB/helmsman/tutor/quasi-mate was preoccupied with orienting the nominal mate to the bridge in strange waters, while handling radio communications and conning the vessel. These duties

to do. The pairing of an experienced AB with a new officer on the *Empress of the North* was not a problem by itself; inexperienced officers can learn a lot from unlicensed crew and should never miss such opportunities, but there should never be any doubt as to who is in charge. The new third mate's lack of preparation, mixed signals regarding his role, and overreach on the part of the AB produced a breakdown in team performance. The *Empress of the North* is a dramatic example of a team built on a weak foundation disintegrating under slight pressure. So let's focus next on how resource management relates to teams.

far exceeded his normal workload and obviously exceeded his abilities. Meanwhile, the third mate was attempting to learn the bridge and settle into his new role under equally trying circumstances. While a two-person bridge team should have been enough, these two people were insufficient for these circumstances.

The Shared Mental Model

A shared mental model is a prerequisite for effective teamwork. Without a unified concept of tactics and goals, a team can only succeed through luck and timely initiative. The chief components of a shared mental model are planning and conferring. The *Empress of the North* bridge team had neither.

Despite having personally devised this ad-hoc bridge team, the captain set forth no expectations to bind them to a common strategy, and the passage plan contained few useful specifics. A shared mental model implies that the team has a clear sense of goals and expectations. Obviously, there was an expectation to keep the ship in safe water, but compounding the lack of guidance were three different graphical representations of the track line. While both the northern and southern routes were navigable, only the southern route made any sense for this bridge team; yet the choice was left open. The significance of leaving this choice open revealed itself in the final seconds before grounding when Rocky Island loomed up out of the darkness, unnerving both men, sowing indecision and drawing the ship in like a magnet.

The best opportunity to establish a shared mental model through conferring occurred at the beginning of the watch. But the AB, who adopted officer-like duties, was not present for the handover. The quality of the handover was further degraded by the comings and goings of other crew members and their unrelated conversations. Due to the level of distraction, the conference that the new third mate held with the second mate would not have been optimal even if he was experienced, and it certainly did not rise to what this situation called for. This, of course, points back to the captain's lack of involvement and forethought. The lack of a shared mental model handicapped this team from the start, but it became especially evident at Rocky Island.

Remember, the goal of a shared mental model is not to get the team to agree for the sake of agreement. In a team setting, the shared mental model exists to help the left hand and the right hand work together.

Zones of Responsibility

On a vessel, members of a team need clear zones of responsibility. These zones may be associated with traditional roles (captain, chief mate, chief engineer, AB, etc.), or they be narrower assignments in the context of a standing team or a mission team. Like puzzle pieces, zones of responsibility should fit neatly together to produce a complete big picture for the team leader. Unlike puzzle pieces, however, a team is better when the zones of responsibility can flex and bend with the situation so that the big picture remains intact in fluid situations. As a member of a team, this flexibility does not mean abandoning your assigned post, switching between tasks, and stepping on toes. But it does mean being willing and able to step outside the bounds of a specific assignment to fill in gaps in the team's situational awareness as needed. On the *Herald of Free Enterprise* the boatswain noticed that the assistant boatswain was not at the controls for bow doors just before sailing, but this situation was unrelated to his own duties so he did not consider sharing that information with anyone. Investigators wrote: "He took a narrow view of his duties and it is most unfortunate that this was his attitude."

There are many opportunities for members of a bridge team to adjust their zones of responsibility in order to support one another. Just because there is a designated lookout doesn't mean other crew can't also scan the water while tending to their primary duties. While one person may be handling radio communications, others can still keep an ear out to help catch a message in a moment of distraction or interpret a garbled transmission. This assistance did not happen on the bridge of the *Balsa 37* in Tampa Bay when it might have made a difference. A person assigned to radar can help another correlate echoes with visual targets, or with the chart, and vice versa. Support can come in the form of a gentle query: "Do you see that vessel?" Or a reminder: "We're coming up on the call-in point." Or an unsolicited statement of fact when it appears likely to be useful information: "The last position puts us right of track, and the sounder shows 16 feet under the keel." The ideal team member is able to reliably cover his or her assigned zone of responsibility while maintaining some awareness of the overall goal. For this kind of overlap to occur, individual workloads must be light enough to permit some mutual monitoring and occasional assistance. This should be part of team design.

On the bridge of the *Empress of the North* there were no zones of responsibility. The AB/helmsman was a quasi-officer, the officer was a quasi-observer, and the lookout was absent. While there was some inevitable overlap of situational awareness, there was no clear team leader. With no clear leadership or delineation of responsibilities, information coming in had no ultimate destination where it would be used to produce a decision.

Team Leadership

It is not enough merely to divide responsibilities. Teams need leaders. Moreover, the team needs to know who the leader is, and the leader needs to know that he or she is the leader.

In the wheelhouse, "I have the conn," or "Do you have the conn?" are simple, professional phrases for clarifying who is in charge. In addition to a routine watch change, this custom can help prevent an overly deferential officer from unconsciously abdicating responsibility each time the captain appears. On the other hand, for the captain who behaves as if his or her mere arrival on the bridge relieves everyone else of any useful role, these words make the transfer of responsibility official. Unspoken transfers of responsibility can lead to unintended consequences; transfers of responsibility should be explicit, not implied.

With leadership comes accountability, which is a powerful incentive for getting things right. With the benefit of this perspective, the leader can assign roles, delegate responsibilities, and coordinate the efforts of others toward the ultimate purpose, while filtering out the critical information from the noncritical. The team leader sets priorities for the team by soliciting information ("What is the CPA with that vessel?"), establishing benchmarks ("I want to maintain a CPA of one mile with that vessel"), and establishing intervals and standards of accuracy ("Get a radar fix every six minutes"). Finally, the team leader, while fallible, plays a necessary role in making final decisions when there are choices.

On the bridge of the *Empress of the North* team leadership was deeply unsettled. During the investigation the third mate was asked, "So of the two of you, you and the helmsman, who did you see as having the greater responsibility and authority?" The third mate replied, "Him." The transcript from the voyage data recorder clearly captures the progression whereby the AB/helmsman, who at first adopted the officer role by initiating the time, place, and rate of the turn, gradually retreated from responsibility as he lost confidence, and the new third mate took the initiative just in time to go aground (see transcript next page).

Shared leadership is unavoidable at times and may even be desirable with a large team and a complex situation. However, the *Empress of the North* is not an example of that ideal, and, in any event, two people cannot lead the team at the same time. The third mate's lack of assertiveness from the outset may have inadvertently encouraged the AB to compensate by taking the initiative in ways that were not helpful and which he would not have otherwise done. Consciously or not, the AB and the new third mate reversed roles, and it led directly to a breakdown of accountability.

Team Communication

Effective bridge teams require a full range of communication skills. This doesn't mean mariners need to be especially eloquent, but they do need to be aware of the basic communication barriers that were discussed in the previous chapter, and methods for overcoming them.

As noted earlier, it takes time to convey information and meaning from one person to another. For the team leader, this means planning further ahead. For the rest of the team, this means trying to anticipate the leader's needs. Sometimes the team leader is too preoccupied to solicit information. Skilled team members watch for this and feed the information to the leader, inserting it into the dialogue in a way that won't be overlooked but doesn't distract or clutter the airwaves. A free flow of information is crucial to effective teamwork, but it must be disciplined and purposeful, not a din of chatter. Effective team members understand that there is a rhythm, pace, and timing to an operation, and communicate accordingly.

A characteristic of a healthy bridge team approach is that it creates the opportunity for more than one mariner to analyze and interpret a situation. This

Bridge Audio Recording from 4 Minutes Before Accident Until 30 Minutes Afterward

Time / Speaker	Transcript
0126:24 AB-1	yeah. I'm kind of comin' on the inside of our turn here comin' around.
0126:27 3M-1	okay. **. flashing green ** yup there it is.
0126:43 AB-1	that is our flashing green isn't it?
0126:43 3M-1	yup.
0126:44 AB-1	way up there.
0126:48 3M-1	yeah broad on the starboard bow.
0126:52 3M-1	we'll pass inside of that.
0126:53 AB-1	yup.
0127:12 3M-1	are these depths? on there are those in feet or they in uh fathoms?
0127:19 AB-1	fathoms.
0127:20 3M-1	oh # got plenty of water.
0128:53 AB-1	comin' around fast enough you think?
0128:55 3M-1	yeah.
0129:06 3M-1	turn it down *.
0129:27 AB-1	I don't think so.
0129:28 3M-1	you don't think so?
0129:30 AB-1	no.
0129:44 AB-1	we should go on that side. whaddyou think?
0129:52 3M-1	yeah let's just steady up on this until we clear that light.
0129:58 3M-1	oh come left.
0129:59 AB-1	comin' left.

Transcript of the conversation between the AB and the new third mate on the *Empress of the North* bridge on the night of the Rocky Island grounding.
(National Transportation Safety Board)

diversity of perspective can increase the chance that, collectively, the correct interpretation will be reached. This is a distinct advantage over the team-of-one, which is more exposed to the single-point error. Without hierarchy, multiple viewpoints might create confusion, but the chain of command addresses this by giving the team leader the authority to make the final decision. On balance, this system works well. But if team members do not put their interpretations forward, then diversity of perspective might as well not exist. The leader whose team members won't speak up remains as exposed to the single-point error as the team-of-one. This is what happened aboard the Coast Guard cutter *Cuyahoga* one clear night on

0130:02 3M-1	hard left.
0130:03 AB-1	hard left.
0130:07 3M-1	yeah we're real close. we better call the captain. we better call the captain.
0130:10 AB-1	oh #.
0130:11 3M-1	call the captain.
0130:13	[sound of impact]
0130:16	[sound similar multiple warning alarms from GE console]
0130:21 ?	want your radio.
0130:23 3M-1	oh God.
0130:25 ?	* captain *.
0130:29 ?	captain.
0130:29 ab-1	oh he knows he's comin'.
0130:30 CPT	what was that? what was that?
0130:30 ?	that was the rocks sir.
0130:32 CPT	the rocks.
0130:36 3M-1	** rocks.
0130:37 CPT	stop.
0130:38 AB-1	stop stop stop.
0130:45 CPT	we got to call the Coast Guard…hold it hold it here.
0130:46 ?	okay.
0130:47 CPT	hold it.
0130:56 UHF-?	report.
0130:59 ?	we just came inside the rocks.

the Chesapeake Bay in 1978 during an officer training cruise.

On October 20, the 125-foot *Cuyahoga* was approaching the Potomac River from the south at 12 knots when the captain saw one red and one white light of a vessel up ahead on the port bow. The target presented a small radar echo and the captain surmised that it was a fishing vessel heading for the mouth of the river and in same general direction as the *Cuyahoga*. However, the officer candidate who had reported the light had seen an additional range light on the vessel that the captain missed, and which indicated that the target was probably a southbound ship on a more or less reciprocal heading. The officer candidate and the captain shared a mental model to the extent there was a vessel to the north, but the number and configuration of the lights, the type of vessel, and the vessel's probable heading were not addressed. A few minutes later the deck watch underwent a transition, and a different officer candidate took the watch. Studying the target, he too saw the additional range light indicating a southbound ship. The lookout reported "a series of lights" without specifying the number or colors, or the fact that they had the appearance of a ship. Knowing that the captain was already aware of the other vessel, the lookout's report was not passed along. The captain asked the quartermaster to get a radar range on the target, which he did, but the captain did not request a radar plot and nobody initiated one. No one presented the alternate interpretation of the target lights to the captain, and, as the two vessels drew closer, the captain firmly believed he was overtaking a slower vessel on a parallel heading.

Since he intended to turn west into the Potomac River, the captain concluded it was best to overtake the other vessel on its port side, so he ordered a large course change to port. However, the target was not a slow north-bound fishing vessel but a 521-foot southbound freighter making over 14 knots. The freighter had not made radio contact because, up until the *Cuyahoga*'s course change, it appeared to be a routine port-to-port meeting. The captain of the *Cuyahoga* had not radioed the other vessel because he had a locked mental model as to what he was looking at. The *Cuyahoga* was run down, dragged though the water by the oncoming ship, and sunk, killing eleven people.

Communication within a navigational team is not the shipboard equivalent of a roundtable discussion. But it should not be a repressed "permission to speak, sir" environment either. The fact that some of the people on the *Cuyahoga* were officers-in-training did not disqualify them from participating in the team. On the contrary, that is what they were there to practice. The possibility that hierarchy interfered with a more functional team in this case is very real. Team communication should be focused on the task and channeled by the chain of command. When a bridge team goes silent, it means one of two things: either everything is going extraordinarily well, or the team is dysfunctional.

*****

Teamwork involves intangibles that are difficult to quantify, just like the other human factors we have discussed. The quality of teamwork is at least partly a reflection of individual competence. Poor skills yield poor teams. But the quality of teamwork is also a reflection of team chemistry and collective intuition, in other words, relationships. This dynamic is occasionally apparent in sports when modest individual statistics seem to contradict the ability of a team to win. Issues of competence aside, there cannot have been much chemistry on the bridge of the *Empress of the North* at the time of the grounding.

The advantage of any team is that it allows the leader to gain access to more resources and greater expertise. This range of perspectives and solutions can increase the likelihood that good decisions will occur; what is difficult or impossible for an individual to accomplish may be quite manageable for a team. But the team's composition must be appropriate and a shared mental model must exist. Team members need clear zones of responsibility, along with a capacity to enhance the situational awareness of the team by adjusting to changing needs. Communication must be clear and proactive and channeled to support the team leader, and the chain of command must be clear. Bridge teams, like all teams, are a paradox. They must be structured enough to be reliable but flexible enough to permit resourcefulness. Conformity is part of teamwork and so is individual initiative when it serves the overall goals of the team.

Chapter Thirteen

Decision Making and Leadership

LEADERSHIP AND TEAMWORK ARE inextricably linked, as are leadership and decision making. Not all decisions are made by leaders, but all leaders must make decisions. Thorough planning and well-designed procedures can help steer us toward good decisions, but the most critical decision making takes place in the unscripted territory beyond plans, checklists, and standardized procedures. Sometimes situations decision making is a deliberative process where information and the merits of different options are carefully sifted and weighed. Under more dynamic conditions decision makers must act quickly, often with incomplete information. Many situations fall somewhere in between—we have time but nowhere near as much as we'd like. Some decisions, are routine, or closely resemble past decisions, whereas others are unfamiliar, making past experience less instructive. Mariners, particularly those in leadership roles, must learn to cope with the full spectrum of decision-making circumstances.

Barriers to Effective Decision Making

Whether deliberative or dynamic, it's easy to point out a bad decision after it has run its course. Unlike night-vision binoculars, there is no technology for picking out poor decisions in advance, nor do the conditions always exist for producing better ones. There are certain conditions that are more likely to produce poor decisions, however, and mariners should know what they are. In his book *Sources of Power*, Gary Klein has a chapter entitled "Why Good People Make Poor Decisions." Klein identifies three sources of poor decision making: lack of experience, lack of information, and ineffective mental simulation, i.e., failure to mentally visualize potential outcomes of decisions before taking action. While Klein's approach is primarily used to describe nonroutine decisions made under dynamic conditions (emergency responses or split-second decisions such as collision avoidance in extremis), it can be useful when looking at other kinds of decisions too.

Lack of Experience

The first source of poor decision making is lack of experience. In the case of the new third mate on the *Empress of the North*, it was clear that although he possessed technical skills and training, he lacked the experience to make good decisions in that situation. The new third mate agreed to take the watch without completing his planned orientation, without meeting the basic requirements for bridge familiarization, and without a viable passage plan. This was a poor decision that had much to do with his inexperience. It was also the captain's decision to place him in that situation, however, which provides a stark example of poor mentorship of a young mariner. But experience does not arrive with the mail; often it comes from coping with the unfamiliar. While the captain's decisions were of greater consequence, the new third mate's choice provides an important lesson. Lives did not depend upon the new third mate stepping up that night. On the contrary, lives depended upon him *not* stepping up. Knowing your own limits is crucial to good decisions.

Inexperience is unavoidable if people are going to advance and accept new responsibilities. At every stage of a career, people must do something they haven't done before and make a decision they haven't made before. One clue that inexperience may be leading toward a poor decision is the perception that you have no choice. Inexperienced people don't see the choices that more experienced people do. This is how the new third mate came to find himself on the bridge that night.

Lack of Information

A second factor in poor decision making is a lack of information. How can people make good decisions if they don't have the facts? The captain of the *Anne Holly* was already underway when he learned that no helper boat was available to assist in transiting the St. Louis bridges. If he had obtained that information while still at the dock he would have had the opportunity to make a more informed decision. Due to poor communication and a failure to monitor the radar and radio, the pilot aboard the *Balsa 37* in Tampa Bay did not know that the tug *Seafarer* was overtaking another unit. If he had had this information it is unlikely he would have found himself in the position of proposing a starboard-to-starboard meeting that had slim chance of success. Incorrect information is a form of missing information, as demonstrated by the faulty position data displayed on the *True North*'s electronic chart system. In all these cases better information was available, it just wasn't applied to the decision-making process.

Planning and standardized procedures are information-gathering systems mariners use to make routine decisions. Nonroutine decisions, like the one the captain of the *Empress of the North* made in putting the new third mate on watch, require extra effort to get the best information and weigh the options. Since incorrect information is always a possibility, when time permits mariners must be prepared to cross-reference and seek other perspectives before making critical decisions. Investigators asked the captain of the *Empress of the North* if he had conferred with others as part of his decision-making process before putting the new third mate on watch. He replied, "No, it was pretty much just, you know, I didn't even talk about it." As we can now see, this was an important decision.

Lack of Imagination

The third source of poor decision making, according to Klein, is inadequate mental simulation; decision makers see danger signs but explain them away or find reasons not to take them seriously. As a mariner, I think of this as a lack of *professional imagination*, the failure to envision the potential consequences of our professional decisions, and failure to ask the reasonable "what ifs." A lack of imagination in this context includes not picturing how a course of action could go wrong. This is what happened to the captain of the *Empress of the North*. A lack of imagination includes failing to see what the situation might look like if crucial assumptions prove incorrect. The locked mental model very much falls into this category, such as we saw aboard the towboat *Mauvilla* on the Mobile River. A stubborn insistence that things will go according to the plan can play a part in this. Perhaps most important, a lack of imagination is what prevents someone from recognizing that a particular decision, while seemingly insignificant, can have enormous ramifications. In the course of his long career the captain of the *Anne Holly* surely contemplated how he might respond to a runaway barge, but he never imagined how his runaway barges could endanger over two thousand lives. If we are going to avoid poor decisions, we have to imagine such things.

A lack of imagination in professional decision making is something we expect in a greenhorn. It is one reason mentorship is so important. Inexperienced people lack a rich store of experiences and examples to draw on; therefore their ability to imagine a wide variety of outcomes is limited. Ironically though, highly experienced people, the people most qualified to imagine a wide array of outcomes, sometimes fail to do so as well. This may speak to the role of complacency in poor decisions.

Lack of experience, information, and imagination do not always work separately; they can join forces. Picture a relatively new vessel operator who is inadvertently provided with incorrect information regarding the air draft of a crane barge that must be transported. Questioning that information might appear somehow confrontational, or too "by the book," or signal an unwillingness to go along with the way things are done. So the operator sails, either with misgivings or misplaced confidence concerning the air draft and ends up striking a bridge. Lack of experience, information, and imagination were all at work. Conversely, an experienced operator moving the same rig may have done so many times before, but one day the water level is different, or the crane has been stowed differently, or clearance has been temporarily reduced by bridge maintenance activities. Declining to check the data, he too hits the bridge. Experience wasn't lacking, but he failed to gather adequate information or to consider alternate scenarios and outcomes.

In addition to Klein's trilogy, we should recognize some other impediments to good decision making. Low situational awareness due to any of the causes discussed earlier, lack of resources, and insufficient time to weigh options can all prevent people from arriving at good decisions.

Throughout this book we have seen instances of experienced people making decisions that had disastrous consequences with little deliberation, even when they had time. While the schedule is a ceaseless source of pressure, few of these poor decisions were made under the extreme pressure of a truly dynamic situation. In most instances, we must conclude, individuals were unwilling or unable to imagine consequences of their decisions.

Anatomy of a Bad Decision

In putting the new third mate on watch without preparation or support, the *Empress of the North*'s captain primarily demonstrated a lack of professional imagination. There are reasons he may not have been at his best: a medical emergency involving a passenger two days earlier had produced considerable stress and had been a significant job-related distraction. The sick passenger had to be taken ashore and later died in the hospital. The captain had averaged a moderately deficient 6 hours of sleep a night over the previous two days. These factors surely contributed to low situational awareness with regard to handling his next dilemma. However, his decision was made 12 hours before the new third mate went on watch, and it was not made with bad information. When asked by investigators why he didn't use an alternate watch arrangement he responded, "I didn't consider it." There were signs that this decision was flawed, but he found ways to explain away their significance.

What were these signs? First, his decision required him to shelve the sound plan approved by his employers to give the new mate a week for orientation. His decision meant putting an officer in charge who had not participated in drills or completed a safety orientation; therefore, in the event of emergency, he was unprepared to carry out his duties. His decision involved shuffling the watches of the ABs while overlooking the fact that the problem, and therefore the solution, had nothing to do with the unlicensed crew. His decision also required him to ignore the example of the first engineer, who rearranged his own schedule so he could stand watch with the new third engineer who had arrived the same day as the new third mate. Finally, the captain's decision required him to not consider what he would do if there was no new third mate to turn to. These were all signs of a bad decision.

The captain's lack of imagination went beyond not seeing the warning signs. It also led him to not weigh other viable options:

- Splitting the watch between other officers, including himself, and giving the dock watches to the new third mate so that the others could get rest.
- Giving the new third mate the watch, but monitoring his performance and coaching him.
- Monitoring the watch handover to be sure the new third mate had a grasp of the bridge equipment, route, procedures, and expectations and had established his authority.
- Providing explicit guidance via night orders and verbal coaching and adding a "Call me" point before the Rocky Island course change.
- Delaying sailing until a qualified person could take the watch.

This last option would have either cost money or put a crimp in the schedule, or both. We can only speculate as to how the home office would have reacted if the captain had pursued this option. But if the grounding had caused the vessel to sink in frigid waters and darkness with 206 passengers on board and loss of life, a delayed sailing would have looked downright responsible.

Default Decisions

Default decisions occur when a course of action is set in motion based on the path of least resistance. Default decisions aren't always bad. A default decision may follow a familiar pattern of other, similar decisions and therefore have a sound basis. But default decisions can also arise around a nonroutine issue if we take too simplistic an approach. The fact is, not all decisions can be handled methodically, nor should they be. Research shows that people who make split-second life-and-death decisions under dynamic conditions (firefighters, emergency medical responders, and soldiers) often cannot recall any decision-making process at all. They default to their training and experience, which is how training and experience are supposed to work: they create a path of least resistance toward a correct action. Mariners also receive training that is reinforced through drills and practice, and by playing out scenarios of how things could go wrong, and what we would do if they did. When situations arise that fit a mental template based on formal training, self-training, or experience, our decisions are, in a sense, predetermined. There is a cognitive efficiency to this, but it does come with some risk that the similarities between two situations may be more obvious than the differences, and the differences may prove crucial. Also, when an unfamiliar situation arises, default decisions may follow a path of least resistance based on *convenience* rather than a rigorous appraisal of potential consequences.

The decision to put the *Empress*'s new third mate on watch was a default decision that followed a path of least resistance away from optimal results. The new third mate's license was as valid as the other third mate's, making his availability an irresistible convenience for the captain. But there was a huge difference in ship-specific knowledge and overall experience that the AB could not make up, and that difference proved crucial. Furthermore, the AB undertook a role that confused the chain of command. The fact that it was a nonroutine situation may have made it harder for the captain to see the flaws in what, for him, was an easy solution.

Unhesitating split-second responses are required in certain professions, and that sometimes includes the marine profession. Mistakes made under those circumstances are judged differently. In the case of the *Empress of the North*, the captain had time and he had options but, as he said, to investigators, "It wasn't really a decision per se." When an important decision appears to follow a path of least resistance, it can be a sign that a poor decision is in the making.

Uncertainty

Uncertainty is part of a decision maker's world. There are times when information is scanty, unreliable, ambiguous, or too complex to make sense of, yet decisions must be made. Many people avoid making decisions until they have to. Indeed, sometimes keeping options open is sensible in a fluid operating environment. A lot of very good captains avoid being pinned down as to their plans partly because they know how much plans can change. They have a lot of experience with uncertainty and don't want to waste effort trying to plan the unplannable. But when uncertainty delays action for too long, windows of opportunity can close and events can render the would-be decision maker irrelevant. This scenario describes the buildup to many collisions, where people wait and defer action until it's too late. An approaching hurricane has a way of doing this too, as we weigh whether to stay or go, and where to go. If we wait too long, however, we may end up with no options at all.

When uncertainty doesn't cause paralysis, it can prompt a person to make a rash decision. The 282-foot passenger schooner *Fantome* disappeared in 1998 in the Caribbean Sea when the captain attempted to skirt the edge of a Category 5 hurricane, leaving no margin for error. When the hurricane did not follow the path predicted by the computer models, the ship was caught, and all thirty-one crew perished. Hurricanes always present uncertainties, and they are very unforgiving. Attempting to outfox a monstrous storm with the slenderest of margins when he could have stayed in port showed decisiveness, but it proved to be a terrible decision.

Because uncertainty can never be entirely eliminated, decision makers must often proceed without all the information they would like. The explorer Samuel Champlain crossed the Atlantic Ocean twenty-seven times between 1599 and 1633. He traversed huge areas of unmapped North America on foot and surveyed vast stretches of the uncharted shores of the (future) United States and Canada. Champlain dealt with uncertainty on a large scale. He used the word *prévoyance* to describe his approach to leadership and decision making. *Prévoyance* has no precise equivalent in English but has been described as an ability to prepare for the unexpected in a world of danger and uncertainty. It involves learning how to make sound judgments on the basis of imperfect knowledge. *Prévoyance* is learned from experience rather than a book, but it is critical to anyone aspiring to be a good leader and decision maker, particularly under dynamic conditions.

Not all poor decisions lead to terrible consequences. This is why dangerous practices often persist long after the hazards are known. And the best decisions possible may still end badly. Shipboard decision making often entails uncertainty, and the best decision makers learn to strike a balance between caution and action. If the captain of the *Empress of the North* had given more consideration to what was at stake and the potential downside of his decision, it is likely he would have made a different plan. A dash of pessimism is the secret ingredient in making safety-critical decisions, as mariners are not paid to hope for the best.

Leadership

We have all seen good and bad leadership and learned from both. For all the leadership theories that have come and gone, the need for it is fundamentally primitive. Leadership is a constant, and nowhere it is more necessary than on ships with their chain of command and, those most archetypal of leaders, captains.

While perhaps the most obvious leader aboard, the captain is not the only source of leadership. Every officer has leadership obligations, even a junior one on the first day of work. It comes with the license. People ashore also have leadership responsibilities, and they profoundly impact what happens aboard. Port captains, port engineers, and other company officials are all leaders, and they have crucial roles in setting expectations and standards and in supporting those who must meet them. In both the *Herald of Free Enterprise* and the *Andrew J. Barberi* cases, shoreside leadership was particularly lacking in ways that directly contributed to those outcomes.

People can find themselves in leadership roles for reasons other than their leadership skills. This may seem backward but reality often imposes other criteria; timing, availability, the right license or endorsement, special skills, experience, local knowledge, seniority, and personal connections can all play a part in filling leadership positions—but they do not necessarily guarantee leadership ability.

Leadership doesn't always come with a title. It may be exhibited at the lowest levels of a hierarchy, reinforcing what some have always suspected, that leadership involves certain intangibles of character that transcend rank, credentials, and connections. But while some people seem to be born under a leadership star, leadership can be learned, improved, and cultivated by example, necessity, or sheer determination. People grow into it all the time.

Like other arenas, the nautical world contains a full spectrum of leadership types. There are some who bully and belittle while others go out of their way to mentor and instill confidence. There are extroverts who narrate every thought, and introverts whose concerns and expectations are a mystery to the crew, keeping everyone slightly off balance. There are good everyday leaders who are affable, easygoing, and well suited to normal operations but who may be out of their depth in a crisis. There are great crisis leaders who are difficult the rest of the time, which is most of the time. There are reluctant leaders who are exceedingly skilled in all things nautical, except in projecting an aura of leadership; people *know* who is in charge, but they don't *feel* it.

Some leaders place great store by getting along, running a happy ship, and having everyone on a first-name basis. Indeed, no one wants a miserable crew, and, make no mistake, looking out for the crew *is* good leadership. But where lives are concerned, there are more important things than being liked, especially if familiarity muddles decision making. When "getting along" jeopardizes readiness and attention to detail, people may end up in the grip of some avoidable crisis, and very unhappy indeed. If a happy ship isn't possible, leaders are still obligated to run a safe ship. It is impossible to measure the misery avoided by doing things right, and the worthlessness of having been liked when things really go wrong. But a leader who is austere, unapproachable, and lacking compassion may command the respect of subordinates but never get the best out of them, and never be told critical information that he or she needs to hear.

Sometimes the best and the worst of leadership traits are combined in the same person. Skilled, charismatic leaders can be distracted by personal issues and private demons, which only goes to show that real-life leaders are as flawed and as gifted as other people, and that successful leadership styles are as varied as personalities. Great leadership is a balancing act. Occasionally someone possesses the whole package. How fortunate. But usually leaders are a bundle of strengths and weaknesses, and they must work to cultivate the strengths while managing their weaknesses.

Designated and Functional Leadership

A distinction is sometimes made between designated leadership and functional leadership. A designated leader derives authority from rank, title, or place in the chain of command. A functional leader is a person others look to for direction regardless of that person's place in the chain of command. The authority of a functional leader may not be official but it is real; it is a natural outgrowth of knowledge, demeanor, and reputation.

Functional leadership that is not designated has an important place aboard. Think of the lifelong AB who orders people away from a line creaking under strain while the new mate is looking elsewhere. Or the grizzled boatswain who calmly applies the windlass brake while a young officer gazes speechlessly at a runaway chain flying out the hawsepipe in a cloud of rust and scale. But functional leadership without designated leadership can also blur the zones of responsibility. Aboard the *Empress of the North* the new third mate was the designated leader but the AB provided functional leadership, at least up to a point. We already know it was a mistake to put the third mate on watch that night, but the error was compounded by the misplaced functional leadership of the AB. Without that arrangement the new third mate would most likely have had the good sense to call the captain earlier, rather than be led along by an unqualified crew member.

Designated leadership comes to an officer through written exams and a license. Functional leadership is achieved through other tests. A leader whose authority is designated but not functional cuts a pathetic figure. Designated leadership is the shoes; functional leadership is what fills them. The ideal officer, therefore, combines functional leadership with designated leadership in the exercise of the full measure of his or her responsibilities.

Leadership Qualities

Describing successful leadership is always tricky. What works in one setting or for one person does not always apply to other people or other situations. To some extent we have to be who we are and do what works for us. Nevertheless, in talking with effective leaders and watching them, certain characteristics recur in all walks of life and can be summarized as follows.

Self-awareness: By definition, leaders influence the people around them. Consciously or not, they toss the stones that make the ripples. Effective leaders are conscious of the impact their words and actions have on the people they lead, both implicitly and explicitly. Yes, it is important to be yourself, but when you are a leader you serve those whom you lead, even if you are the one giving orders.

Situational awareness: Leaders must maintain the big picture as appropriate to their position. This means avoiding the things that degrade situational awareness and embracing opportunities to improve it. Without adequate situational awareness a leader cannot manage resources effectively or protect the people he or she is responsible for.

Interpersonal skills: Effective leadership depends on strong interpersonal skills—listening, explaining, resolving conflict, and building trust. These are sometimes referred to as soft skills, yet for many people they are hard. Communication in all its forms is immensely important to good leadership.

Motivation: Leaders need to get things done, not ponder the possibilities. Effective leaders are motivated and are able to motivate others. Being energetic and positive helps with this, but motivational style comes in all shapes and sizes.

Respect: Perhaps the most universal trait of an effective leader is the ability to maintain the respect of those he or she leads. Leaders earn respect—i.e., are held in high regard—by exercising the qualities discussed above and, of course, through mastery of their nautical responsibilities. A mariner who lacks sufficient knowledge and skill will not command respect on the water.

Leadership Techniques

Some disciplines draw a bright line between management and leadership. On a vessel, however, leaders must manage and managers must lead, particularly in a small crew. This is true not only in watchkeeping but in the discharge of many duties. In addition to the qualities and traits outlined above, there are a number of specific leadership and management techniques that effective leaders frequently apply in their daily interactions.

Lead by example: It's an oldie but a goody. Because a leader causes things to happen, he or she is always being watched. Personal and professional conduct, work ethic, communication style, and outward attitude are important parts of a leadership persona. By providing positive examples of these qualities, a leader earns respect and trust while modeling expectations. Poor examples in these same areas can leave a designated leader unable to provide functional leadership.

Leading by example does not always mean "leading from the front," to quote another leadership mantra. When a passage plan needs to be researched the mate should not be out proving that he knows how to operate a needle gun, though the occasional spell "in the trenches" can set an example like no other.

Set expectations: People like to know where they stand and what the standard is. They cannot meet expectations if there is no target. Leaders must consider time and resource constraints when setting expectations, and prioritize accordingly. A leader who doesn't do these things cannot avoid blame when things go wrong. Communication skills are a necessity in setting expectations. Articulating expectations takes time and energy, but failing to deliver this aspect of leadership is a failure to lead. The captain of the *Empress of the North* could have done much more for the new third mate in this area.

Delegate: Nobody can do it all alone. That is why there is a crew and a chain of command in the first place. A leader must have the wisdom to delegate in order to preserve situational awareness. Delegating is not for personal convenience; it is for the good of the ship and must be based on realistic expectations. While people often rise to the trust placed in them, leaders must determine which responsibilities to delegate and which to keep for themselves. If a person is given more than they can handle, the fault lies with the person in charge. Delegation was ineffectively handled aboard the *Empress of the North*.

Provide oversight: Even when expectations have been articulated and responsibility appropriately delegated, an effective leader must keep a finger on the pulse. There is a fine line between responsible oversight and micromanaging, though. If a leader must tread this line, it takes only one incident to learn that it is better to err on the side of too much oversight than too little. However, micromanaging has its own problems, and a single letdown cannot become an eternal excuse to withhold all trust from other professionals. Chronic micromanaging can undermine teamwork and morale, but a lack of oversight is a lack of leadership. The captain of the *Empress of the North* did not provide effective oversight of the situation.

Shift perspective: As discussed, shipboard leaders are expected to maintain a perspective that is appropriate to their place in the scheme of things. A mate is not expected to have the perspective of a company owner, but there is a big picture appropriate to being a mate, and it usually borrows a little from the perspectives above and below the mate in the chain of command. A leader who is caught up in minutiae can lose perspective, leaving the entire operation exposed, but a leader must also have a sound grasp of the details in order to make good decisions. The ability to shift easily from the big picture to a small detail, and back again, and adopt different perspectives is crucial in maintaining situational awareness. This applies in the short-term to tasks such as voyage monitoring and collision avoidance, as well as to long-term management decisions.

Be assertive: Leaders must not hesitate to assert themselves when acting in their leadership capacity. This includes not permitting practices that run counter to their professional judgment. It can also include declining tasks for which adequate resources are not available, or for which the individual is not prepared.

Be decisive: Leaders are expected to make decisions, especially in difficult situations. In difficult situations, more than ever, subordinates look for signs of decisiveness, and sometimes for signs of indecision. But leaders also need to recognize when to reverse a bad decision. Clinging to a poor decision for appearance's sake only makes matters worse, and waffling is bad too, so leaders must be prepared to back down from a bad decision as soon as they realize what is happening.

Apply emotional intelligence: Emotional intelligence is the capacity to understand and communicate your own emotions, and to understand and appreciate the emotions of others. It plays a crucial role in dealing with everyday people problems. Emotional intelligence is directly related to the development of the brain's frontal lobe in late adolescence, where key executive functions such as organization, planning, impulse control, and decision making occur.

Emotional intelligence is not psycho-babble. It is one of the most important drivers of professional success in adults. Eighty-one percent of the competencies that distinguish outstanding leaders from average ones are related to emotional intelligence. Emotional intelligence has been found to be roughly *twice* as important as IQ in distinguishing top performers from average ones. Emotional intelligence is the ability to put yourself in the shoes of someone who is putting himself in your shoes. Effective leaders tap into this resource in their daily interactions, whether they realize it or not.

Be ethical: Ethical behavior is a widely recognized prerequisite for effective leadership. It is an important ingredient in commanding respect and in building a good reputation. Ethical behavior entails obeying the law, but it embraces a more organic notion of right and wrong even when there is no explicit guidance. The maritime world is a highly regulated field, and mariners don't always agree with the regulations. However, consciously violating them is unethical and is usually a slippery slope. Falsifying logbooks, as was done aboard the *Herald of Free Enterprise*, is a classic example of unethical behavior. Concealing a medical condition, as the assistant captain did on the *Andrew J. Barberi*, was a seemingly private deception but it ended up affecting a great many people. Among other problems, unethical behavior sets an example that may be followed by others in ways that have unforeseeable consequences. Ethical behavior means doing the right thing, not because you have no choice, but when you do have a choice and no one is watching. Ethical behavior is not a bid for sainthood; the world is full of imperfect people who conduct themselves ethically.

Project command presence: A function of leadership is to inspire others to have confidence in themselves and in their leaders. This is accomplished in many ways but almost always requires projecting an unharried presence. Command presence is established by way of verbal and nonverbal cues that convey beyond a shadow of a doubt who is in charge. In reality, people in leadership positions don't always feel calm, confident, or in control, but they have a duty to try and act it. For those without much leadership experience, it is often in the trying and the acting that they find they can do it, and that people will follow them.

Among smaller crews an atmosphere of informality often prevails, but this doesn't necessarily mean that the chain of command is weak or that there is a lack of appropriate command presence. However, a bit of shipboard formality can provide an atmosphere conducive to the exercise of command presence.

The Leadership Persona

A leadership persona is the delivery system for all the other qualities of leadership. It is how you come across when you are in charge. It includes technical proficiency, communication style, command presence, and a host of subliminal messages communicated through dress, body language, habits, idiosyncrasies, and more.

A leadership persona is a blend of innate personality traits and a more deliberate concept of the leader you want to be. Though it is normal to have differences between the professional self and the private self, a successful leadership persona cannot be alien to the leader's nature. An introvert cannot pretend to be an extrovert for long, and vice versa. Since no one can prescribe the correct leadership persona for another, you must cultivate your own, while understanding that the way you come across shapes the way other people respond to your leadership.

How do you create an effective leadership persona? First, look around and observe other leaders. By rejecting traits and techniques that appear ineffective or feel forced, and by adopting those that strike you as effective and that resonate personally, you can begin to construct an authentic leadership persona that is true to your nature and wears well for everyday use. As you gain experience, your leadership persona evolves, and eventually it is completely natural and someone is emulating *you*. Vague as this may sound, it is exactly what many fine leaders have done without an ounce of leadership training, though some have been more purposeful in their approach than others.

Second, be self-aware. By taking stock of personal strengths and weaknesses, a person can build on strengths and compensate for weaknesses. True objectivity is impossible, but a reasonably self-aware person can improve him- or herself. In June of 1944, Major Richard Winters landed on the beaches of Normandy and successfully led a company of paratroopers across Europe through some of the fiercest combat of World War II. Winters was not a career soldier, but his achievements are often studied as examples of how an individual can rise to the responsibilities of leadership under exceptional circumstances. He characterized the development of his own leadership persona this way: "If you take advantage of opportunities for self-reflection, and honestly look at yourself, you will be able to be a better leader."

Chapter Fourteen

Human Error

To err is human. Despite plans, procedures, systems, checklists, alarms, experience, and boundless good intentions, errors are made every day, and all mariners make them. Most errors are insignificant. Either a mistake is caught in time or it simply doesn't produce noteworthy consequences: a watchkeeper misses a course change but brings the vessel back on track without incident. Sometimes we are aware of a serious brush with danger and breathe a sigh of relief at our luck; other times there might be minor damage. We should learn what we can from such incidents because they are a bargain.

There are also *ghost errors*, the close shaves we're not aware of: a shoal narrowly missed and never spotted; a misplotted position that doesn't matter because you were relying on radar at that point; a misspoken order that turned out all right because it was misunderstood; transposed numbers that subsequent events rendered immaterial; an incorrect calculation that another error canceled out; a false assumption that had no consequences, and no one was the wiser. No one tracks these errors, but they surely happen.

And then there are the errors that the world gets to read about, such as when the captain of the tug *Mercury* tore open a petroleum barge in Puget Sound because he believed that confirming his position at regular intervals was unnecessary. His theory had been "proved" by personal experience—until the fundamentals of watchkeeping caught up with him. Let's be clear about two things: a lack of consequences is not the same as good seamanship, and people often escape consequences for reasons they can't take any credit for.

The point of examining marine accidents is not simply to avoid repeating someone else's specific mistake. While that is desirable, it is far better if we can learn to view our professional surroundings as a world in which similar errors and accidents are constantly in the making and are constantly being averted, either by luck or design. Good judgment comes from experience, and, unfortunately, experience often comes from bad judgment. The cases we have examined here have expanded our inventory of vicarious experience, and thus enhance our ability to recognize accidents in the making.

Types of Human Errors

J. T. Reason, a British psychologist, has provided a useful way of classifying human errors. He divides them into two groups, active and latent. *Active failures* are unsafe acts that occur on the front lines of an operation: a watchkeeper neglects to confirm his position, plan the route, monitor the weather, or follow procedures, leading directly to an accident. *Latent conditions* are part of the system in which an individual functions. Unsafe latent conditions set the stage for active failures and make them more likely to occur. The time and place of specific active failures are difficult to predict, but if the latent unsafe conditions that make active failures more likely can be identified, the system can be changed before something bad happens. Let's apply this to two of the cases we have examined.

On the *Herald of Free Enterprise*, the assistant boatswain did not report to duty at the bow doors when summoned, and the first mate left the scene without confirming that the doors were closed. These were active failures—individuals did not do what they should have done. However, these active errors were fostered by latent conditions: the long-standing practice of leaving de facto responsibility for the doors to the assistant boatswain, and a procedure that required the first mate to be on the bridge prior to sailing even though he was loading cars up to the last moment. The tide made it necessary to trim the ship for loading, yet there was no provision made for restoring the vessel to an even keel before sailing. Sooner or later, this ship would sail with the bow doors open, trimmed by the head.

On the day the *Andrew J. Barberi* struck the pier on Staten Island, the lookout had left the bridge before being relieved so he could unlash a malfunctioning door to allow passengers to disembark. Meanwhile, the senior captain was doing paperwork in another part of the ship and was not on the bridge. Both of these were active failures since they directly weakened the bridge team. As a result, no one noticed when the assistant captain became incapacitated at the controls. Though the operating procedures aboard the *Andrew J. Barberi* did not anticipate this specific event, they did recognize that a large passenger ferry on New York Harbor warranted a fully functional bridge team. Lax enforcement and inconsistency in applying the procedures were latent conditions in the system that facilitated the active failures.

Latent unsafe conditions wear camouflage; they may be discernible to an expert or a fresh pair of eyes, but they are often invisible to everyday eyes. Sometimes, even after latent conditions have been exposed, people resist changing practices because they are invested in the status quo. "It has never been a problem in the past" is a common reaction. By definition, a latent condition never is a problem for the past, only for the future. The irony is that new crew, who may have the clearest view of latent errors in a shipboard system, may also have the least standing to effect change and the most to lose from "rocking the boat." The window of time, upon joining a vessel, to recognize latent conditions is quite small and it closes as the newcomer accepts things as they are. Some latent conditions have surprisingly simple solutions, while others do not. Some solutions will win support, while others may not. Improving any system has costs and benefits, and each individual must weigh what is at stake.

Both active failures and latent conditions involve *errors of omission*, something that was not done but should have been, and *errors of commission*, something that was done that should not have been. Errors of omission are far more common than errors of commission. Mariners are far more likely to skip a step, forget a detail, neglect to confirm, become distracted, or fail to call the captain than to turn to port when they meant to turn to starboard. If we are interested in detecting human errors, latent unsafe conditions and errors of omission are good places to start.

Error Chains

Single-point errors are a genuine concern in bridge resource management, as we have seen. But usually it requires a convergence of errors and events to produce an accident. Some accidents result from purely human acts (failure to confirm the vessel's position or repair a latch), but they also can be aggravated by conditions beyond individual control (fog or severe weather). The term *error chain* describes how seemingly insignificant and unrelated ordinary acts, events, and behaviors can link up to produce a catastrophe. With an error chain, an accident is really the final stage of a progression that has been building for hours, days, or even years. Each link in this progression brings the accident closer. In hindsight, error chains are easy to trace, however shocking the result may seem.

After the captain of the *Maria Asumpta* sailed onto the rocks in Padstow, Cornwall, with the loss of the ship and three lives, many people defended him, saying he was an outstanding seaman, a consummate sailor, and a professional beyond reproach who had experienced momentary bad luck when the engine failed at a critical moment. Yet it is clear that an error chain was set in motion with his inadequate passage plan that he altered for the worse at several points. Each decision to sail closer to shore with an unfavorable wind and current was a link in a chain that ended with his vessel in a dangerous position by choice and despite warnings. The failure to foster an effective bridge team was a more subtle error, but it may well have contributed to the foundering of the vessel.

All the cases we have explored involved a traceable error chain, even those triggered by the unlikeliest of events. Error chains can be broken, however, by chain spotting or foreshadowing.

Chain Spotting

Chain spotting is the simple act of noticing that something is awry and intervening, possibly in the nick of time. Chain spotting relies on situational awareness, and it is a constant duty of a watchkeeper. The distracted mate on the *Queen of the North* had 13 minutes to realize he had missed the waypoint and was heading toward land. The error chain could have been broken if he had glanced at the radar, if either member of the bridge team had noticed land visually, if VTS had noticed the vessel deviating from its route, or if one of the other watchstanders had returned to the bridge in time to jar the bridge team's attention from their personal concerns. A last-ditch attempt to break the chain failed because the helmsperson did not have a thorough understanding of the new steering system, another link in the chain.

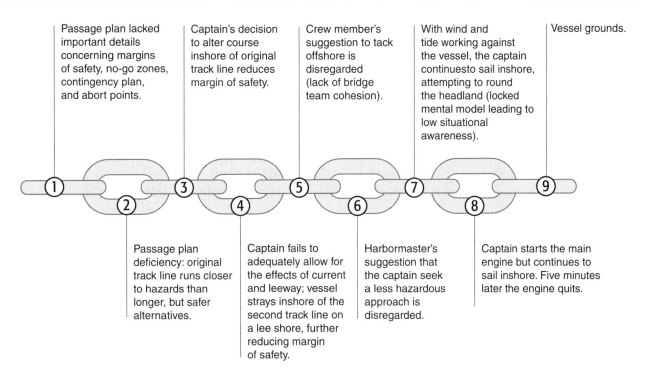

A depiction of the error chain that led to the grounding of the *Maria Asumpta*.

Chain spotting is one of the reasons watchkeepers have jobs. Error chains detected in time don't make it into the papers, which means watchkeepers have done their jobs. But because chain spotting is reactive rather than proactive, successful chain spotting does not necessarily mean all is well. If errors continue to be spotted around the same issue, that is a clue to a latent condition, an accident waiting to happen. Chronic chain spotting suggests that a system-wide fix is in order, and this requires someone to step up and take a leadership position on the issue.

Foreshadowing

Whereas chain spotting involves catching things going wrong, foreshadowing is projecting the current situation into the future to assess the *potential* for things to go wrong down the road. It is sensing possibilities in advance, rather than simply swerving to avoid catastrophe. Foreshadowing is rooted in past experience, either firsthand or vicarious: a situation reminds you of something you have seen before, heard about, or were taught. The appraisal stage of passage planning calls for a degree of foreshadowing regarding the potential hazards of a route, but the value of foreshadowing extends to all aspects of vessel operations. Foreshadowing is an essential part of every successful mariner's methodology, for it is the means by which hindsight is exchanged for foresight, or *prévoyance*. Let's see how the failure to foreshadow led to two of the incidents we have discussed.

The captain of the *Safari Spirit* committed the most ordinary of errors when he miscalculated a clearing distance, losing his vessel on the very hazard he had set out to avoid. All calculations and measurements are subject to such error—thus the old adage "measure twice, cut once." The ordinariness of some tasks can lead to complacency, reinforcing the need for mariners to double-check their own work and the work of others around them. If the captain had considered the potential for a mistake in his work and projected the possible consequences into the near future, he may have used more care or multiple references, or both.

The three vessels in the Tampa Bay accident converged in darkness at a bend in a narrow channel. Two mates, two pilots, and three captains failed to see the potential for misunderstanding or miscalculation despite their accumulated expertise. Given the law of averages, a collision is always unlikely, and this fact surely influenced the thinking of all parties. But in this case the potential was greater, and no one took steps to compensate for that fact. Everyone was paying attention, but no one acted on the possibilities. Complacency is an error chain's best friend, as it makes light of our better instincts.

Situational Awareness

Both chain spotting and foreshadowing are forms of accident avoidance that rely on situational awareness as well as individual initiative. If mariners are alert and focused they are more likely to notice things going wrong. If they have a grasp of the big picture they are more able to project events into the future with greater accuracy. Managing factors that degrade situational awareness—overreliance, distraction, stress, fatigue, complacency, and transition—puts them in a better position to recognize error chains in motion and take action to prevent accidents. Experience does not guarantee a lifetime supply of situational awareness, but knowledge gained through things going wrong does nurture the so-called sixth sense, which is a powerful tool in breaking error chains. Just as the words "if only" are stamped on every link of an error chain, the words "what if" are often part of breaking one.

The Swiss Cheese Model

Another construct for how accidents happen is called the Swiss cheese model. This concept portrays our defenses against accidents as a series of Swiss cheese slices set on edge.

Since no single line of defense is invincible, the barriers invariably have holes of varying size and location. The holes in our defenses against human error are not static; they move and change in size with the strength of the resources in the system and the conditions under which we operate. When the holes in our defenses align, a "trajectory of accident opportunity," as J. T. Reason puts it, is created that can allow hazards to pass through all the defenses and cause the entire system to fail. An underlying assumption of this model is that accidents are the result of flawed systems, rather than unsafe individuals, therefore

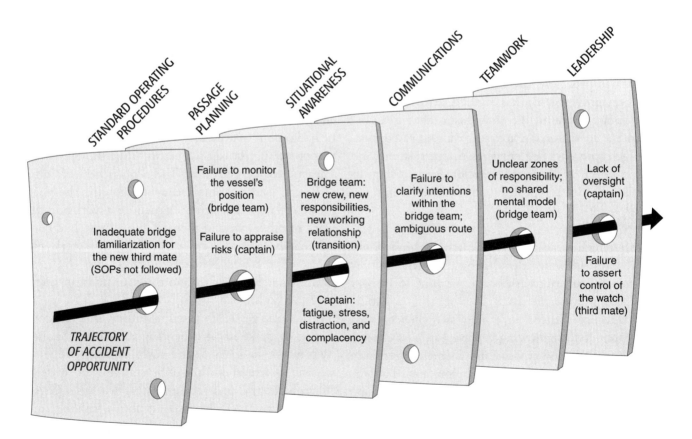

A depiction of how the Swiss Cheese Model applies to the *Empress of the North* grounding. The slices represent defenses against human error but in this case they contained holes that aligned, producing a trajectory of accident opportunity.

defenses against human error in the Swiss cheese model emphasize system-wide solutions over individual action.

For instance, a watch rotation designed to produce a well-rested crew can provide an effective defense against the memory lapses and inattention associated with fatigue. Under normal conditions, this slice of cheese might be almost completely intact. But after a stretch of rough weather, an emergency, or protracted operations, fatigue may increase to the point that holes appear in that particular line of defense. In another example, an organization may take measures to produce greater crew continuity, such as assigning crew to the same vessel, or at least to the same class of vessel, in order to avoid accidents associated with high crew turnover (transition). But some crew turnover is unavoidable, which will periodically produce holes in that defense. The existence of one or more holes does not necessarily mean an accident is inevitable or that a system is deeply flawed, but as the holes become larger and more numerous, the possibilities for alignment increase, and an accident becomes more likely. Maritime accidents do not come out of blue; they only seem to. Understanding this is critical to accident prevention.

Error Trapping

Breaking error chains is good; trapping them is better. A timely warning to someone that a companionway ladder is slippery can prevent that individual from getting hurt. Renewing the nonskid treads, painting the steps a bright color, illuminating the ladder, and putting up a warning sign are all error traps. Now, even without that timely warning, many people can avoid injury.

Though accidents and close calls may leave individuals wiser, if the conditions that produce them persist they will recur, perhaps in a slightly different fashion and with more severe consequences. Error trapping is the process of examining the context and root causes of an accident or a potential accident and changing practices to prevent future accidents of the same nature. Error trapping fits well with the Swiss cheese model, as it amounts to neutralizing latent conditions by improving lines of defense against recognized problems.

Error trapping can be as basic as installing a depth-sounder where it can be easily seen or locating a radar unit so a watchkeeper can see the screen and the chart at the same time, while also facing forward. Or it can be complex and far reaching. Load lines, drug testing, and double-hulled tankers are all examples of large scale error traps, and all met with opposition at the time. Error trapping may involve some combination of improved procedures, training, or equipment, but at the heart of it is a fundamental recognition that all people are fallible, and that fallibility is not necessarily negligence.

Unfortunately, people are better at error trapping after something has gone wrong rather than before it. After losing a vessel due to a simplistic approach to passage planning, the company that operated the *Safari Spirit* instituted a more rigorous approach to passage planning. In the wake of the *Andrew J. Barberi* accident, bridge procedures were overhauled and all deck officers in the fleet underwent specialized BRM training. After the *Mauvilla* caused the Sunset Limited tragedy, radar training became mandatory for inland vessel operators, and a task force was created to assess the vulnerability of bridges spanning the nation's waterways. BRM training itself is a form of error trapping by drawing attention to the human component in safe vessel operation.

Some errors are more obvious than others, and some are harder or more expensive to trap. Attempts at error trapping often cause debates over whether the right error is being trapped the right way or if a problem ever existed in the first place. To be sure, error traps are imperfect, and they do not always have the desired effect. Error traps constructed in the wake of an incident may benefit those who come after, but who wants to be the one whose misfortune inspires the next error trap? It is always best to do your error trapping preemptively if possible.

Building Error Traps

How do you know when an error trap is needed? Sometimes you don't, and that is one reason why we will never live in an accident-free world. But minor accidents, near misses, or a pattern relating to either one may suggest a need for a better system. For instance, vessels in the *Herald of Free Enterprise* fleet had sailed with their bow doors open on at least five previous occasions. Several people had recognized the problem and advocated that bow-door indicator lights be installed. This equipment was not required, and the matter was not pursued. Within days of the disaster, all the ferries in the fleet had bow-door indicator lights.

Nine months before the *Scandia* ran aground, an engine-room fire occurred while it was dockside. The crew extinguished the blaze with equipment borrowed from a local fire department. Recognizing the significance of the episode, the captain requested air packs, spare tanks, and turnout gear. It was a commonsense error trap. This equipment was not required, however, and the matter was not pursued. Nine months later the lack of equipment aligned with other "holes"; another fire broke out, the tug burned and grounded, causing a major oil spill. As professional mariners, we must not confuse what is required with what is needed.

Creating a safer vessel operation system is often a battle against other priorities and entrenched practices. Some organizations have recognized the long-term benefits of error trapping by encouraging the reporting of active failures and latent unsafe conditions without fear of reprisal. At a minimum, successful error trapping also requires the following conditions.

Awareness: There must be awareness of the hazard, either by personal observation or by sharing information. Debriefing incidents (and drills) are powerful tools for raising awareness and should be part of normal operations.

Analysis: There must be an opportunity for informed analysis of close calls and existing procedures. Too often either the expertise is not available or other priorities take precedence before any useful reflection can occur.

Resources: Whatever form an error trap takes, there must be adequate resources to implement it. There is no point in mandating a safer way of doing things if time is not provided to carry it out.

Adaptation: The lightbulb really does have to want to change, and this may be the toughest part. If mariners can't or won't change what they are doing, then unsafe conditions will remain latent until a bigger accident really shakes things up. Both shipboard and shoreside leaders have a role in fostering a culture of collaboration around producing safer practices.

Accident Cascades

Sometimes an unbroken error chain (or the sudden alignment of holes in defenses) leads to an *accident cascade*. Like a gust of wind scattering campfire sparks across a dry meadow, an accident cascade can spread and build momentum at a rate that overwhelms all efforts to control events. Like dock lines parting under strain, as one defense collapses, the strain increases on those that remain.

An accident cascade is often triggered by seemingly minor events that serve as catalysts for bigger events. Chasing down latent conditions is often beyond the scope of a mariner's daily duties, so that all that stands between an accident cascade and a typical day's work may be an individual who sweats the small stuff even as complacency beckons. While effective leadership demands a comprehensive vision, it is attention to detail—sometimes tiny details—that can keep a ship safe on a day when resolving latent conditions is on someone else's list, and nowhere near the top of it. Attention to detail, like chain spotting, is not a long-term solution to structural problems, but as a short-term survival mechanism, it may be even better than a PFD.

The Tug *Valour*, 2006

Strictly speaking, BRM is meant to train watchkeepers in the human component of the job. All other things being equal, effective planning and procedures, high situational awareness, and productive human interaction create better results than not having those things. We began this book by examining the sinking of the *Herald of Free Enterprise*, an incident that had almost nothing to do with navigation or watchkeeping in a conventional sense, yet spoke volumes about bridge resource management. We close now with a case that, similarly, has little to do with navigation but everything to do with how experienced people on a good ship can thoroughly undermine themselves and their vessel by not applying the principles of sound bridge resource management.

On January 15, 2006, the oceangoing tug *Valour* sailed from Delaware Bay bound for Texas pushing a barge with 175,000 barrels of oil. Fuel and water tanks had been topped off and slops discharged. There was no liquid ballast and no partially full (or slack) tanks at the time of departure. The tug was 125 feet long with a beam of 36 feet.

The *Valour* was an uninspected oceangoing tug with a load line certificate and a stability letter that outlined specific operating requirements pertaining to stability.

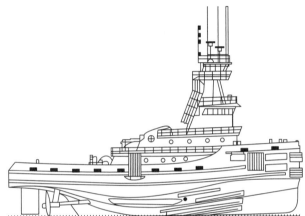

The offshore tug *Valour* before the accident.
(United States Coast Guard)

Like virtually all stability letters, this one stated that "any cross-connections between port and starboard tank pairs shall be closed at all times when underway." This instruction was to prevent a list from developing or being exacerbated by liquid flowing from one side of the vessel to the other. Like other stability letters, the *Valour*'s also instructed the captain to "determine the cause of any list before taking corrective action." This is sound advice for several reasons. High weights, off-center weights, and the free-surface effect can all contribute to a vessel's list, but each involves different corrective action. The wrong action can make matters worse. Stability letters are standard operating procedures of the highest order. Although these operating instructions are addressed directly to the captain, all deck officers bear responsibility for maintaining stability, and engineering officers are obliged to inform the captain of noncompliance issues found in the engine room. If a mariner is inclined to adopt lax operating standards, the stability letter is not the place to do it as there are no minor stability casualties.

The *Valour* was built and maintained for offshore work, and its nine crew members had impressive levels of experience. Seven crew members had more than twenty-five years at sea, including the captain, and five had more than thirty years. Most of the crew, including the captain, had been with the *Valour* for two years or more. While experience and continuity were not lacking, there were holes in this outwardly robust profile.

For one, the tank information provided by the operating company stated that the #18 ballast tanks (empty upon departure) each had a capacity of about 8,500 gallons when in fact they could hold nearly twice that. Second, there was no way to gauge the contents of these tanks since there were no sounding tables or tank level indicators for them. Ballast transfers were based on elapsed time using a 250-gallon-per-minute ballast pump. But since the #18 tank capacities were incorrect and there stated, it was not possible to know the status of these tanks when they were between full and empty. Information is a resource, and this lack of information was significant.

A third concern was that the vessel had a slight permanent list to starboard. Although a permanent list is best corrected with permanent ballast, this was not done. Instead, if an officer chose to, he could temporarily correct the list by pumping ballast into the port #18 tank until the vessel was on an even keel. This practice, of course, created a slack tank each time it occurred. Though the permanent starboard list was mentioned in the stability letter, and the assistant engineer, who had been aboard for two months, knew about it, the captain did not. The captain also did not know the pumping rate of the ballast pump, though he used it to conduct ballast operations based on elapsed time. As fuel and water were consumed, lists were corrected with liquid ballast, but the chief mate had formed a habit of correcting very minor lists of only a few degrees with liquid ballast. In addition, on at least one occasion the chief engineer conducted ballast operations on his own initiative without informing anyone until afterward. When told about it, the captain responded that this action was acceptable though not normal protocol. There was no company policy on ballasting, nor did the captain have his own policy. Two other company captains interviewed later stated that

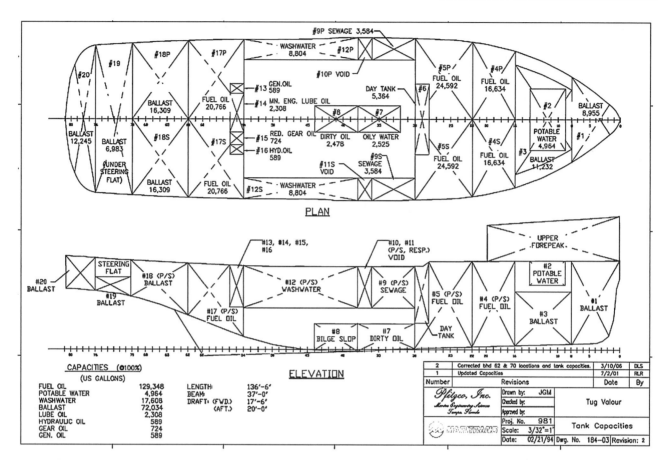

Tank #18 was aft of the fuel tanks. The Coast Guard received this corrected tank capacity plan after the *Valour* sank. (United States Coast Guard)

they did not conduct, or allow, ballasting operations to correct minor lists. They also stated that they did not allow anyone but themselves to authorize ballast operations. Under the *Valour*'s captain, ballast operations were not tightly controlled.

There was also a contradiction between the stability letter instructions about keeping the port and starboard cross-connections closed "at all times when underway" and actual shipboard practice. The captain believed that the cross-connections could be left open in protected waters, but the letter did not say that. Moreover, there was no evidence that cross-connections were ever managed based on the vessel's location. On the contrary, it was standard practice to cross-connect the washwater tanks and the tank pairs that fed the day tank. The assistant engineer knew what the stability letter said about the cross-connections, but he accepted the shipboard practice, citing that it was widespread throughout the fleet. It is not known what the chief mate or the chief engineer thought of this practice.

Lastly, although the second mate had thirty years of seagoing experience and had held a master's license for two years, he still considered himself a rookie and was not confident standing watch alone under conditions that he considered out of the ordinary. The captain dealt with this by camping out on the wheelhouse settee as needed.

On January 17, two days after departure, the *Valour* was southbound about 40 miles off North Carolina. In anticipation of severe weather the tug moved out of the barge's notch and towed the barge astern. By 1500 the wind had built to 25 to 35 knots. The chief mate noticed a slight port list in the tug that he corrected by pumping 15 minutes of ballast into the starboard #18 ballast tank. At 1930 the captain came on watch and pumped it out because the vessel had adopted a slight starboard list. The #18 tanks were both now empty.

At 2130 the second mate relieved the captain. The wind was now 50 to 60 knots, gusting to 70, with 15- to 20-foot seas out of the south. As the second mate

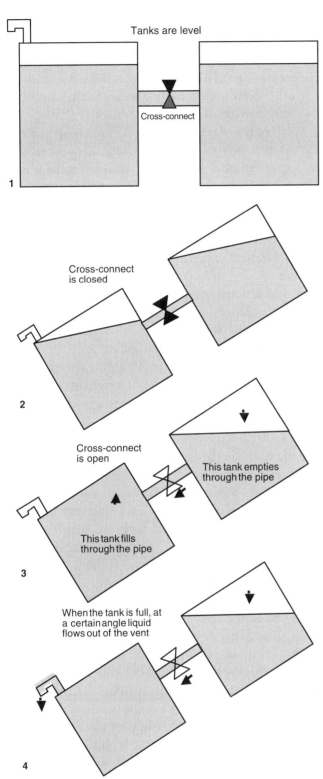

Four possible cross-connection scenarios. The tanks are level when the boat is level and the cross-connection between them is closed (1). The *Valour* is heeled over, but with the cross-connection closed the weight does not move from one side of the vessel to other (2). With the cross-connection open, the lower tank fills (3). Once the lower tank is full, at certain angles fluid flows out of the tank overflow vent (4). (United States Coast Guard)

was not comfortable being on watch alone, the captain remained on the bridge. The tug was now carrying its usual slight starboard list, and the chief mate came to the bridge at 2200 and remarked on it. The second mate had never given ballast orders before but since the captain was on hand, the captain agreed to deal with it. The captain called the engine room and ordered the chief engineer to pump 15 minutes of ballast into the port #18 ballast tank to counteract the list. According to the Coast Guard report, both the captain and the second mate understood at this point that the captain was in charge of the ballast operation, although the chief engineer was not aware of this. The second mate later told investigators that he was unaware of the requirements in the stability letter.

We do not know why, but after 15 minutes had elapsed the chief engineer continued to pump. We can only speculate that, from his perspective, the list had not yet been removed so he did what he thought best. After 45 minutes had gone by, he called the wheelhouse and the second mate picked up. The chief engineer asked if the vessel was now on an even keel. The second mate said it was not. He did not put the captain on the phone, he did not inform the captain that the chief engineer was still pumping, and he did not question the fact that the chief engineer was still pumping. The captain, just a few feet away, did not learn that the chief engineer had far exceeded his orders, and, since no one told the chief engineer to stop pumping, he continued to pump ballast into a tank that was nearly twice the size of its stated capacity.

Sometime between 2300 and 2315 the vessel leveled out but then began to list noticeably to port. Shortly afterward, the captain called the chief engineer to ask him what was going on. He then ordered the chief engineer to pump out all ballast, but his confidence in the engineer was now shaken so he took the conn from the second mate and sent him to the engine room to get a report. The second mate did not tell the captain of his phone conversation with the chief engineer, or that ballast had been pumping for over an hour.

By 2320 the vessel had developed a 15-degree port list. Greatly concerned, the captain sounded the general alarm. He told the crew over the public address system to get their life jackets and survival suits. Several of the crew didn't hear him or didn't follow his directions. When the second mate reached the engine room he saw the chief engineer standing between the main engines. Noise precluded conversation but the chief engineer gave him the "okay" sign. Precisely what was "okay" is unclear, but the second mate returned to the wheelhouse and reported it to the captain.

Still concerned, the captain again called the chief engineer to find out what was going on. Not satisfied with the response, he again sent the second mate below to the engine room. This visit was also inconclusive. The chief engineer said something the second mate couldn't hear and pointed to an area aft of the port main engine. The second mate stood beside the chief engineer and saw some water sloshing in the bilges but it was not an excessive amount. The second mate relayed the information to the captain, still failing to mention the phone conversation with the chief engineer, despite all that was now going wrong.

Down in the engine room, the chief engineer told the assistant engineer to check if the cross-connection for the #17 fuel tanks back aft was closed. It was. The chief engineer then went to the forward part of the engine room. The second engineer assumed that the chief engineer was checking the #4 port and starboard fuel tanks to make sure they were not cross-connected. He also assumed that the cross-connections between the #5 port and starboard fuel tanks were open as these tanks were currently feeding the day tank. All we know for certain about this is that divers later found the #4 and #5 fuel tank pairs and the wash water tanks with their cross-connections open. This would have allowed liquids to flow from the high side to the low side of the vessel, increasing the list that the ballast transfer had induced.

At 2334, with the *Valour* now listing 25 degrees to port, the captain issued a Mayday call. Though he still didn't know the cause of the list, at least two of his officers (the chief engineer and the second mate) knew that the #18 ballast tank had been overfilled, and it seems likely that at least one engineer knew that there were open cross-connections between multiple tank pairs. The cascade was just beginning.

On his way to retrieve his survival gear, the chief mate slipped on a companionway ladder, apparently breaking both his legs and going into cardiac arrest. The second mate, now second in command, responded to the medical emergency, leaving the captain alone to manage the vessel, the tow, the communications, and the rest of the emergency. Some others attempted to assist with the injured chief mate, but at 2338 one of the ABs slipped and fell overboard. The captain immediately ordered another AB to grab some marker lights from the emergency locker but the AB thought the captain meant signal flares. Not finding any signal flares, he departed the bridge of his own accord to help with the injured chief mate without informing the captain of his actions. Meanwhile, two other crew were grappling with a refrigerator that threatened to break loose.

At 2345 the chief mate stopped breathing and the second mate began CPR. The *Valour* was now rolling 30 degrees to port and the AB was still in the water. Though the captain had designated someone to keep an eye on a green light in the water marking the AB's location, no MOB apparatus had been deployed.

The chief engineer had continued to empty the port #18 ballast tank as ordered, but now he and the captain decided to stop discharging from the port #18 tank and start pumping seawater into the starboard #18 tank in an effort to counterbalance the port list. Many mariners would concede that stability is not their strong suit, but you do not have to be an expert to see problems with this plan. Pumping seawater into the starboard #18 tank would add weight aft when the stern was already awash. Adding weight to the high side could adversely affect the center of gravity, particularly under the dynamic storm conditions. Pumping into the starboard ballast tank would produce another slack tank where there had not been one. Another reason it was a poor strategy was that it committed the pump to an unachievable task since the sea suction for filling the starboard tank was nearly out of the water at this point due to the list. Whereas emptying the port tank was feasible and sensible, filling the starboard tank was not. But accident cascades do not always promote rational thinking, and, as we know, the terms of the stability letter were not well understood aboard this boat.

Though the captain disagreed later, there is some evidence that in this interval the vessel got onto an easterly heading. If true, this would have placed the full force of the weather on the starboard side of the *Valour*, heeling it farther to port and intensifying the flow of tank liquids from the high side to the low side. Hampered by fatigue, distractions, and unthinkable stress from the loss of two crew, attempting to stem a full-blown accident cascade almost single-handedly, it is conceivable that the captain lost situational awareness with regard to his heading.

Just before midnight the captain became concerned that the barge might trip the tug, so he summoned the second mate to the wheelhouse to help

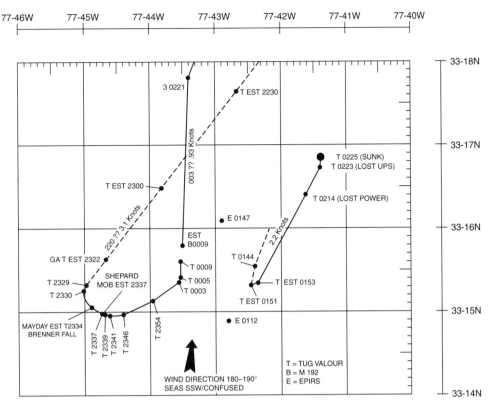

The *Valour's* track line during the accident. (United States Coast Guard)

locate it. Though one of the ABs had already seen that the barge was just off the port quarter of the tug, he hadn't reported it. Upon spotting it, the captain ordered the second mate to go release it. The second mate explained that he did not know how to release it, but the captain sent him anyway. By this time the towing wire was tending well to port, submerging the port bulwarks and permitting seawater to enter vents along the port side. Redirecting the second mate at this stage meant disrupting efforts to revive the chief mate, a fact that can only have increased stress on all parties. With the guidance of an AB, the second mate managed to pay out all the wire, but it fetched up at the end, leaving the tug still tethered to the barge. The second mate notified the captain of the situation and went back to assist the chief mate, who had not had a pulse for perhaps 15 minutes. The captain called the second engineer out of the engine room, and he was able to release the barge.

While a shipmate can never be faulted for striving to save the life of another, we cannot ignore the fact that the second mate's situational awareness does not seem to have kept pace with the circumstances. He was now second in command, and the captain needed all the support he could get. Many lives now depended on coordinated action. Instead, as is often the case, the second mate focused on the one thing he felt he could do well, administering first aid and CPR amid the chaos, instead of delegating it.

Shortly after midnight, the captain contacted the Coast Guard again: "Right now the boat is floating and it looks good." The facts did not support this optimism. The captain was able to maneuver the tug close to the man overboard, and the crew made several attempts to rescue him. The AB was able to retrieve one of the ring buoys, but the crew could not get him aboard.

By 0100 the tug *Justine Foss* had arrived on scene and was standing by to offer assistance. At 0106 a Coast Guard helicopter rescued the lost AB from the water. However, the *Valour's* life raft then broke free and drifted away. The helicopter dropped another life raft to the crew, but it was running low on fuel and could not remain to assist further. The second raft lay alongside for about 30 seconds but blew away too. At this point the captain ordered the second mate to cease efforts to revive the chief mate and told the crew to muster on the bow with life jackets and survival suits. Only about half the crew had survival suits, though there were enough on board for all. The chief engineer was a large man and there was no suit to fit him. The second engineer had a suit but it was the wrong size.

In keeping with the maxim that you never leave the ship until the ship leaves you, the crew huddled on the bow. But what were they waiting for? Efforts to right the vessel had ceased, the engine room was under water, the situation was worsening, and a vessel was standing by to rescue them. At 0223, with the tug trimmed severely by the stern, the bow abruptly shot into the air. One AB fell into the water, and the chief engineer and another AB fell with full force onto the superstructure of the house before sliding into the sea. The second mate was also swept into the water. The first AB was last seen facedown in the water. About an hour later the unresponsive chief engineer was pulled from the water and shortly thereafter went into cardiac arrest and died. The others were rescued, the *Valour* sank, and the accident cascade ended.

The *Valour* was a proven vessel with an experienced crew. Yet several factors created a long error chain. The incorrect information concerning tank capacities and sounding data made it impossible to conduct ballast operations from a position of certainty. Standard operating procedures regarding those ballast operations were absent. As a result, lax communication and blurred zones of responsibility regarding a critical aspect of vessel safety became the norm. The fact that several licensed officers tolerated regular violations of the stability letter suggests that camaraderie and crew continuity did not produce a strong or ethical team when it came to these requirements. The investigation further concluded that the captain's "lack of command presence with his crew and poor knowledge of his vessel exacerbated the environment for this casualty to occur and progress. As a result of his easy-going manner with his crew, he is the primary source of the breakdown of bridge resource management that occurred." The second mate's lack of confidence also deprived the captain of a resource he should have had. What would it take to combine these latent conditions to cause an accident? As is so often the case, the sea itself stood ready to merge seemingly inconsequential acts and omissions into an unimagined tragedy. But the sea did not act alone; it had the collaboration of many people. With a different crew and a different way of doing things, this accident would not have happened.

Failure to Break an Error Chain

As we have seen throughout this book, the things that move events toward an accident can range from the trivial to the spectacular. But most of the clues to the development of an error chain involve latent unsafe conditions. Disparate as these stories are, there are patterns behind error chains, so let's review the clues.

No plan or procedure: Predetermined passage plans and standard operating procedures are vital tools for preventing human error. They serve as memory aids and discourage spontaneous shortcuts. They help mariners maintain situational awareness by bringing structure to the core watchkeeping practice of comparing what *should be* happening to what *is* happening. When there is no plan or procedure for a critical operation, mariners are more exposed to the formation of an error chain. Aboard the *Valour*, neither the captain nor the company believed ballasting procedures were necessary. The result was that people did what they wanted and the situation got out of control. There can't be a policy, procedure, or plan for every situation, but in this case, they could have used one. Experience is the best tool in the box when events overrun plans and procedures, but experienced people make mistakes too. Plans and procedures are not just for new people.

Departure from a plan or an SOP: There is little point in having plans and procedures if they are not followed. There are many reasons this can happen, but when a sound plan or procedure is abandoned, it is a clue that an error chain is in process, even if the outcome isn't immediately obvious. Violating established CPAs, under-keel clearances, margins of safety, rules, or regulations, and skipping an item on a checklist are all clues that an accident could be in the making. Complacency often paves the way for departure from plans and procedures.

The captain of the *Mercury* had been given clear expectations for position fixing (every half hour), but he did not appreciate the value of half-hourly fixes in maintaining situational awareness, so he ignored the procedure that could have best kept him out of trouble. Sooner or later, seat-of-the-pants navigation begets seat-of-the-pants results.

Sometimes unforeseeable events force a sudden change of plan. When the senior third mate on the *Empress of the North* fell ill, the captain suspended key orientation procedures for the new third mate, providing a crucial link in the resulting error chain.

Sometimes a departure from procedures is part of a gradual undermining of standards, in which many people participate. On the *Valour*, not heeding or enforcing the terms of the stability letter was part of a wider culture. That culture did not hold up well when active failures merged with latent conditions.

If you want to spot an error chain or anticipate an alignment of holes in the Swiss cheese, watch out for times when plans or procedures are not followed. If there is something wrong with the procedure, then change it so that it accomplishes what it is supposed to accomplish. Don't sail with flawed, broken, or disregarded procedures.

Ambiguity: Ambiguity is fertile ground for indecision and poor decisions. This is one reason why the COLREGS instruct mariners to avoid subtle course changes when maneuvering to avoid another vessel. Instead they instruct mariners to make clear and decisive course changes so there is no mistaking what the vessel is doing. Some ambiguity is unavoidable, and flexibility is a survival skill. Good decision-makers learn to handle ambiguity when it cannot be eliminated, but tolerating gray areas where none should exist is asking for trouble.

Ambiguity thrives in the absence of plans, procedures, or guidance. The captain of the *Maria Asumpta* courted ambiguity by keeping his options open right down to the wire. His choices for approaching port ranged from the most safe to the least safe, but he never committed to any of them. One by one, circumstances and decisions eliminated his options, finally leaving him with only the worst choice. A better approach to passage planning and monitoring would have removed such ambiguity by separating the good choices from the bad ones in a deliberate, controlled manner. The new third mate on the *Empress of the North* was given minimal guidance and two different routes around Rocky Island. In light of his complete lack of familiarity with the situation, this ambiguity created an error-rich situation.

Ambiguity often arises around communication. This is why professional communication, both internal and external, requires putting forth the very best effort, every day. In the three-way collision on Tampa Bay, incomplete or ambiguous statements led to the formation of an error chain. The failure of more than one person to clarify or request clarification led to a colossal misunderstanding. Because of ambiguous lines of communication on the *Valour*, the second mate failed to relay critical information between the chief engineer and the captain during the ballast transfer. His errors of omission generated mistrust between the engine room and the wheelhouse that led to wasted effort and further misunderstanding, which perpetuated the error chain. Ambiguity in communication is all too familiar to anyone who has ever uttered the words "I thought you were going to do that."

Ambiguity in the chain of command is a problem. Uncertainty regarding who is calling the shots is a clear signal that an error chain is forming. The captain of the *Empress of the North* contributed to ambiguity by constructing a bridge team with unclear zones of responsibility. The new third mate aggravated the situation by acting like a bystander and letting the helmsman make the radio calls and take the vessel where he pleased. Had the captain come to the bridge at the change of the watch and said to the new third mate, in the presence of the AB, "You're in charge. Call me if in doubt." it might have made all the difference. Aboard the *Valour*, the easygoing command style of the captain led to a casual and decentralized approach to ballasting operations. Not all shipboard operations require tight control, but some do. The lack of clarity regarding who was in charge turned fatal when the captain assumed control of the ballasting, while the second mate retained nominal control over the watch and the chief engineer ignored instructions.

In all aspects of vessel operations, no matter how informal the work environment or how familiar the work relationship is, ambiguity has the potential to initiate and perpetuate error chains.

Incompatible goals: Incompatible goals contribute to ambiguity by blurring priorities. Aboard the *Herald of Free Enterprise*, shoreside management's goal for on-time departures pitted searediness against the schedule. The captain of the *Empress of the North* wanted adequate experience on the bridge that night, but he didn't want to disrupt the routine of anyone who actually had adequate experience. The captain of the *Anne Holly* wanted a helper boat to transit the St. Louis bridges, but he also didn't want to wait for one. As events prove, sometimes you can't have it both ways.

It isn't difficult to confuse priorities when trying to reconcile incompatible goals or trying to please two masters, such as safety and the schedule. When mariners find themselves attempting to have it both ways, it should be a clue that a larger problem may be lurking. One response is to seek some objectivity, either by stepping back and reassessing the situation or by getting a second opinion, or both.

Improvisation: The loss of key resources (people, equipment, or information) forces improvisation. Improvisation is a point of pride among mariners everywhere, and rightly so. However, when they improvise essential resources, it may be a clue that they are contributing to an error chain. Mariners should view the "duct tape reflex" with suspicion, especially as a routine response.

People: Insufficient training, knowledge, or experience for a given situation often results in the need to improvise. The illness of the regular third mate on the *Empress of the North* deprived the captain of an important resource, and he attempted to improvise with the new third mate.

Equipment: You never know when you'll need an anchor. That's why lots of vessels that seldom anchor still carry one. The captain of the *Scandia* sailed with an improvised anchoring system for the fuel barge because the windlass was broken. The improvised system did not work when needed, and the absence of functioning equipment may have even spurred his attempt to outrun the storm. When the captain of the *Anne Holly* learned that no assist boats were available, he decided to make do with what he had.

Information: The *True North* had forward-scanning sonar, but, for fear of damaging it, the captain deprived himself of the information it could provide and instead based his situational awareness on other instruments. In the unsurveyed waters of northwestern Australia, depth information had special significance, but he chose to get by without the very information that could have alerted him to the ECS problem.

There are plenty of times when mariners must be willing and able to improvise. But when critical resources are lacking, and they know it, it is time to consider the potential for an error chain to grow out of that situation.

Fatigue, stress, distraction, and complacency: If mariners are exhausted, stressed, distracted, or complacent, situational awareness and performance will suffer. Unfortunately, the nature of all four conditions is that mariners are often unaware when they are being affected by them. To the extent they can be alert to these human performance factors, mariners should realize that they can contribute to the formation of an error chain.

Transition: A transition is a weak point in the most cleverly designed defense. In the world of boats transitions are necessary, and can often be anticipated. Transitions provide rest, fresh perspective, a break in the monotony, and renewed focus by generating new problems to solve, new factors to consider, and new chemistry among the crew. Yet for all these benefits, transitions are also points where mariners can lose situational awareness while adjusting to new circumstances. For a deck officer, whose work is influenced by the natural environment, transitions have particular significance with regard to weather, visibility, currents, and so forth.

Many of the accidents discussed here occurred close to or during transitions of one type or another. In the three-way collision in Tampa Bay, the arrival of a pilot aboard the *Balsa 37* coincided with the captain leaving the bridge, the lookout being stood down, the chief mate relieving the third mate on the bridge, and a failure to construct a bridge team from the remaining crew. Aboard the *Empress of the North* the new third mate's first experience as a watchkeeper was a particularly bleak example of a mishandled transition. Since transitions can contribute to lower situational awareness, they can contribute to the formation of an error chain.

Speed and momentum: On boats, there is sometimes an inverse relationship between speed of advance and situational awareness. The faster information comes at us, the harder it can be to process it. The pilot on the towboat *Mauvilla* wasn't moving fast, but it was faster than he could handle under the circumstances and he became disoriented in the fog. If the only way a mariner can maintain a grasp of events is to slow down or stop, then that is the right speed for the situation.

It is not only speed in knots that can contribute to an error chain. Sometimes an overriding sense of momentum drives people along, contrary to good sense and clear facts. This concept can be applied to many activities, but, in terms of watchkeeping, it is evident that momentum affected mariners' judgment in many of the situations we have examined. Mariners are not paid to go backward or stand still. The point is to move forward, and this inevitably generates a sense of momentum. However, mariners must try to recognize when momentum is driving events, not good seamanship.

Small tolerances: Small tolerances are always a red flag, whether they involve CPAs, UKCs, or taking a minimalist approach to some other aspect of an operation. What constitutes "small" varies widely and depends on many things. Small CPAs are commonplace in some kinds of work, making it difficult to view them as remarkable. Some vessels must drag their skegs through mud to reach a berth—they have no UKC. By themselves, small tolerances are not necessarily a hazard. It is only when something unexpected or unintended occurs that they are a problem. When tolerances are small, so are margins of safety. A mechanical problem, such as occurred on the *Maria Asumpta*, or a simple misunderstanding, such as in the Tampa Bay incident, illustrates what can happen when operating with little to spare. Small tolerances are unavoidable in many kinds of work, but since they expose us to the full force of active failures and latent conditions, mariners should be mindful of their role in error chains.

Adverse environmental factors: From a watchkeeping perspective, perhaps nothing is more important to maintaining situational awareness than a clear view of what is going on. Yet operating in darkness or restricted visibility comes with the territory. Watchkeepers are expected to do it, and be good at it. Even in familiar waters, operating in darkness can be challenging. Depth perception is hampered, and backscatter from shore lights interferes with identifying objects on the water. At night, many visual cues for estimating speed and turning rate are missing, and it takes longer for the eye to detect changes in the vessel's relationship to its surroundings. There are fewer options for cross-referencing. All of these factors were at work the night the *Anne Holly* got underway in St. Louis. Fog and precipitation make things worse. Whatever the source, reduced visibility is an automatic challenge for watchkeepers. To a lesser extent this principle applies when visibility is blocked by land, piers, buildings, anchored ships, etc. If you can't see what's going on, your ability to know what is going on is reduced. Two-thirds of the case studies we've examined occurred during darkness or some state of restricted visibility.

Sometimes mariners can avoid heavy weather, and when they can, they should. The *Scandia* could have and should have, but that was not a reasonable option for the *Valour*. Moreover, the *Valour* appeared well equipped to handle it that night. High river stages, strong currents, ice, and other environmental factors are part of the marine world. What mariners cannot work around they must work through, recognizing that such conditions make it that much easier for an error chain to link up.

*****

Mariners live and work in an imperfect world, and human error is part of it. If we can learn to see the clues around us, then we have an opportunity to replace hindsight with foresight. If we can develop the leadership and communication skills to act on what we know about human error, then that foresight has real value.

Coda: Understanding and Implementing Bridge Resource Management

Understanding and implementing sound Bridge Resource Management is not the same as learning other shipboard skills like navigation, radar operation, or line handling. Those skills, once learned, become permanent and quickly accessible. You may neglect to fully apply that knowledge sometimes, or you may become rusty with certain aspects of it, but the knowledge is still there and can be summoned. BRM is more difficult to call up as a discrete skill. Knowing which aspects of BRM are needed, and how to deploy them, depends on the opportunities that arise and the ability of the mariner to perceive those opportunities.

To compound these intricacies, BRM may appear to contain contradictions. It emphasizes the need for a shared mental model in vessel operations, yet it warns against the hazards of a locked mental model. It emphasizes the role of plans and procedures, while recognizing that people must be able to flex and adapt to changing circumstances. BRM involves making maximum use of resources, but not overrelying on any one of them. BRM recognizes that there is no substitute for experience but that experience can also lead to complacency. BRM advocates good communication, which requires people to speak up, but not in a way that is confusing or distracting. BRM recognizes that qualified crew must be willing and able to improvise, but that a pattern of improvisation around critical safety functions is a clue to the formation of an error chain. Professional team members are expected to show initiative, independent thinking, and raise concerns when something isn't right but also respect the chain of command and put the goals of the team first. BRM recognizes that there are times to be skeptical of conventional wisdom, but also times to trust.

Above all, BRM recognizes that for all our technology, people still count more, and that people are fallible in predictable ways. That accidents will occur is a mathematical certainty. That the correct application of BRM principles can prevent some of them is also certain. By itself, BRM is not a blueprint for success; but when combined with the rest of our professional training and experience, it provides a wider horizon for seeing how things go wrong on vessels, and how to defend against human error.

Appendix

Sample Master-Pilot Information Exchange (MPX)

The following information should be filled in on a card (or spreadsheet) to be handed to the pilot upon boarding.

Ship Name:_____

Present Port:_____

Boarding Point & Destination:_____

Master's Name:_____

Ship Call Sign:_____

Official Number:_____

Home Port:_____

Place and Year Built:_____

LOA (Length Overall):_____

LBP (Length between
Perpendiculars):_____

Distance Bow to Bridge/
Manifold/Stern:_____

Height Keel to Mast Top and
Present Air Draft:_____

Draft/Trim:_____

Increase in Draft for
One-Degree List:_____

Maneuvering RPM/
Speed Table:_____

Time Full Ahead to
Full Astern:_____

Crash Stop Distances:_____

Turning Circle
Information:_____

Blind Spots:_____

Invisible Distance Diagrams:_____

Squat Effect on Under
Keel Clearance:_____

Maximum Rudder Angle:_____

Type of Propulsion and
Horsepower:_____

Bridge Control of Main Engine:_____

Astern Power Percent of
Ahead Power:_____

Astern Engine Operating
Limitations:_____

Starting Air Limitations:_____

Critical RPM:_____

Bow Thruster and Horsepower: _____

Compass Error:_____

Navigation Equipment
Information:_____

Anchor Information:_____

Dangerous Cargo Aboard:_____

Signatures:_____
(Pilot/Captain)
Date:_____ Time: _____

Sources

Prologue
Department of Transport (U.K.). Report of the Court No. 8074, M/V *Herald of Free Enterprise*. Formal Investigation. London, 1987.

Chapter One
International Chamber of Shipping. *Bridge Procedures Guide*, Fourth Edition. London: Marisec Publications, 2007.
National Transportation Safety Board. *Allision of Staten Island Ferry* Andrew J. Barberi. Marine Accident Report NTSB/MAR-05/01. Washington, D.C., 2005.

Chapter Two
International Maritime Organization. *International Convention on Standards of Training, Certification, and Watchkeeping*. 1978.
———. *Guidelines for Voyage Planning*, Annex 24 and 25. IMO Resolution A.893(21). 1999.
———. *International Convention for the Safety of Life at Sea*. 1974.
National Transportation Safety Board. *Fire Aboard the Tug* Scandia *and the Subsequent Grounding of the Tug and the Tank Barge* North Cape. Marine Accident Report NTSB/MAR-98/03. Washington, D.C., 1998.
———. *Grounding and Sinking of U.S. Passenger Vessel* Safari Spirit. Marine Accident Brief NTSB/MAB-04/01. Washington, D.C., 2004.
U.S. Government. *Code of Federal Regulations*. Title 33, Part 96. Washington, D.C.

Chapter Three
Marine Accident Investigation Branch. *Summary Report of an Investigation into the Grounding and Foundering of* Maria Asumpta. January 1996.
National Transportation Safety Board. *Ramming of the Eads Bridge by Barges in Tow of the* Anne Holly *with Subsequent Ramming and Near Breakaway of the* President Casino *on the Admiral*. Marine Accident Report NTSB/MAR-00/01. Washington, D.C., 2000.
Washington State Office of Marine Safety, *Prevention Bulletin 96-02, The Barge 101*, Publication 00-08-008, Revision March 2000.

Chapter Four
Hecht/Berking/Büttengenbach/Jonas/Alexander. *The Electronic Chart, Second Edition*. Lemmer, Netherlands: Geomares Publishing, 2009.
Smith, A. J. *Bridge Team Management*, Second Edition. London: The Nautical Institute, 2004.

Chapter Five
Australian Transportation Safety Bureau. *Independent Investigation into the Grounding of the Passenger Vessel* True North. Marine Safety Investigation No. 205. April 2005.

Chapter Six
Crenshaw, Dave. *The Myth of Multitasking: How "Doing it All" Gets Nothing Done*. San Francisco: Jossey-Bass, 2008.
Professional Mariner, #142, December–January 2011.
Transportation Safety Board of Canada. *Striking and Subsequent Sinking [of] Passenger and Vehicle Ferry* Queen of the North. Marine Investigation Report M06W0052. March 2006.
Richtel, Matt. "In Study, Texting Lifts Crash Risk by Large Margin." *New York Times*, July 27, 2009.

Chapter Seven
National Transportation Safety Board. *Derailment of Amtrak Train No. 2 on the CSXT Big Bayou Canot Bridge*. Railroad-Marine Accident Report. NTSB/RAR-94/01. Washington, D.C., 1994.
Carey, Benedict. "In Battle, Hunches Prove to Be Valuable." *New York Times*, July 27, 2009.

Chapter Eight
Australian Transportation Safety Bureau. *Independent Investigation into the Grounding of the Passenger Vessel* True North. Marine Safety Investigation No. 205. April 2005.
Drew, Dawson, and Kathryn Reid. "Fatigue, Alcohol and Performance Impairment." *Nature*, July-August 1997.
National Transportation Safety Board. *Collision between Passenger/Car Ferries M/V* North Star *and M/V* Cape Henlopen. Washington, D.C., 1988.
———. *Evaluation of U.S. Department of Transportation Efforts in the 1990s to Address Operator Fatigue*. Safety Report NTSB/SR-99/01. Washington, D.C., May 1999.
Parliament of the Commonwealth of Australia. *Beyond the Midnight Oil: An Inquiry into Managing Fatigue in Transport*. House of Representatives, Standing Committee on Communications, Transport and the Arts, October 2000.
Report on the Investigation of the Grounding and Loss of the Cypriot-registered General Cargo Ship Jambo *off Summer Island, West Coast of Scotland, 29 June, 2003*. Marine Accident Investigation Branch, Report No. 27/2003. December 2003.

Sasaki, M, Y. Kurosaki, A. Mori, and S. Endo. "Patterns of Sleep-Wakefulness Before and After Transmeridian Flight In Commercial Airline Pilots." *Crew Factors in Flight Operations: IV. Sleep and Wakefulness in International Aircrews*, R.C. Graeber (ed.), NASA Technical Memorandum No. 88231, Moffett Field, CA: Ames Research Center, 1986.

Smith, Andy, Paul Allen, and Emma Wadsworth. *Seafarer Fatigue: The Cardiff Research Programme*. Centre for Occupational and Health Psychology, Cardiff University, November 2006.

U.S. Coast Guard. *Crew Endurance Management Practices: A Guide for Maritime Operations*. Washington, D.C.: U.S. Coast Guard Research and Development Center, January 2003.

Chapter Nine

Argyris, Chris, Robert Putnam, and Diana McLain Smith. *Action Science: Concepts, Methods, and Skills for Research and Intervention*. San Francisco: Jossey-Bass, 1985.

Higbee, Gary A. "Workplace Complacency: Techniques." American Society of Safety Engineers Development Conference and Exposition, Seattle: June 11–14, 2006.

Senge, Peter, Art Kleiner, Charlotte Roberts, Richard Ross, and Bryan Smith. *The Fifth Discipline Fieldbook: Strategies and Tools for Building a Learning Organization*. New York: Random House, 1994.

Chapter Ten

Chiarizia, Barbara. "Dangerous Assumptions: Three Vessels Collide." *U.S. Coast Guard Proceedings*, Winter 2007–08.

National Transportation Safety Board. *Annual Report to Congress*. NTSB/SPC-99/01. Washington, D.C., 2008.

———. *Marine Accident Brief* Cowslip and Ever Grade. MAB-99/01. Washington, D.C.: 1999.

Professional Mariner, #4, December–January 1994.

Chapter Eleven

Adams, Michael. *Shipboard Bridge Resource Management*. Eastport, Maine: Nor'easter Press, 2006.

Faulkner, William. *Requiem for a Nun*. New York: Random House, 1951.

Chapter Twelve

Diestel, Capt. Hans-Hermann. "Towards Bridge Team Management." *Seaways*, December 2004.

Hutchins, Edwin. *Cognition in the Wild*. Cambridge, Massachusetts: MIT Press, 1995.

Janis, Irving L. *Victims of Groupthink: A Psychological Study of Foreign-Policy Decisions and Fiascoes*. Boston: Houghton Mifflin, 1972.

National Transportation Safety Board. *Grounding of U.S. Passenger Vessel* Empress of the North. Marine Accident Report NTSB MAR-08/02. Washington, D.C., 2008.

Perrow, Charles. *Normal Accidents: Living with High-Risk Technologies*. New York: Basic Books, 1984.

Safety Digest. Marine Accident Investigation Branch, Dept. of Transport, London, January 2005.

U.S. Coast Guard. *USCGC Cuyahoga, MV Santa Cruz II (Argentine); Collision in Chesapeake Bay*. Marine Casualty Report USCG 16732/92368. Washington, D.C., 1979.

Chapter Thirteen

Boyatzis, R. E. "Competencies as a Behavioral Approach to Emotional Intelligence." *Journal of Management Development*, 28: 749–770, 2009.

Boyatzis, R. E. and F. Ratti. "Emotional, Social, and Cognitive Intelligence Competencies Distinguishing Effective Italian Managers and Leaders In a Private Company and Cooperatives." *Journal of Management Development*. 28: 821–838, 2009.

Chen, W. and R. Jacobs. *Competence Study*. Boston: Hay/McBer, 1997.

Fischer, David Hackett. *Champlain's Dream: The European Founding of North America*. New York: Simon and Schuster, 2008.

Goleman, D., Boyatzis, R. E., and A. McKee. *Primal Leadership: Realizing the Power of Emotional Intelligence*. Boston: Harvard Business School Press, 2002.

Klein, Gary. *Sources of Power: How People Make Decisions*. Cambridge, Massachusetts: MIT Press, 1999.

Chapter Fourteen

Adams, Michael. *Shipboard Bridge Resource Management*. Eastport, Maine: Nor'easter Press, 2006.

Reason, James T. "Human Error: Models and Management." *British Medical Journal*, March 2000.

Reddington, Krista. "Communication Breakdown: Failed Assumptions Lead to a Fatal Sinking at Sea." *U.S. Coast Guard Proceedings*, Summer 2010.

U.S. Coast Guard. *Investigation into the Circumstances Surrounding the Sinking of the Tug* Valour. Marine Information for Safety and Law Enforcement. MISLE/2569550. Washington, D.C., 2008.

Acknowledgments

The author would like to thank the following organizations and individuals for their support, assistance, and expertise in the making of this book:

Maine Maritime Academy, the National Transportation Safety Board (U.S.A.), the United States Coast Guard, the Australian Transportation Safety Bureau, the Transportation Safety Board of Canada, and the Marine Accident Investigation Branch (U.K.);

Kimberly H. Parrott, Capt. G. Andy Chase, Capt. J. Samuel Teel, Capt. George Sandberg, Capt. Rob Rustchak, Capt. Robert Jones, Chuck Barbee, Ken Potter, Greg Chaffey, Barry Strauch Ph.D, Jana Price Ph.D, Caroline Hudson, Jon Eaton, Bob Holtzman, Janet Robbins, Margaret Cook, and Molly Mulhern. Special thanks to Sam and Kessler Parrott.

Index

Numbers in **bold** indicate pages with illustrations

A
abort points and radius, **44**, 45, 52–53
accidents and incidents. *See also* errors
    accident cascades, 150, 154–56
    BRM and, 4
    distraction and, xi
    fatigue and, xi, 88–90
    human factors and, xi, 1–2, 4
    marine environment and, 149
    responses to, 149
    single-point errors, 6, 24, 134–36
    situational awareness and, xi
    stress and, xi
    transition and, 100–102, **101**, 111–13, 158–59
active failures, 145, 146, 150
acute stress, 84, 92
adenosine triphosphate (ATP), 91
Admiral, **28**, **29**, 97–98
AGD (Australian Geodetic Datum) 66 chart datum, 69
AGD (Australian Geodetic Datum) 84 chart datum, 69
aids to navigation (navaids), 43
alarms and indicators
    as distractions, 75, 79
    importance of, 79
    turning off or reconfiguring of, 75, 79
Alaska, 123–28, **124**, **126**
alcohol consumption and blood alcohol levels, 89, 90
alternate routes, 53
ambiguity, 126, 157
anchors and anchoring
    anchor ball, inference about, 96
    functional system, sailing without, 18, 20, 21, 158
anchor sleep, 91
Andrew J. Barberi
    accident and SOPs on, 8–11, **9**, **10**, 16
    at-risk behavior and bad luck, 95
    communication devices on, 114
    complacency and, 97
    distraction on, 78
    error trapping after incident with, 149
    fatigue and, 89
    ISM Code and, 12
    latent conditions aboard, 145–46
    leadership on, 141
    unethical behavior on, 144
    written communication on, 115

Anne Holly
    abort point, 53
    accident and passage planning, 26–30, **27**, **28**, **29**
    complacency and, 95, 97–98
    conferring and decision making, 31, 32, 138
    error chain aboard, 158, 159
    getting underway transition, 106
appraisal stage of passage planning, 18–21
A. Regina, 88–89
Argyris, Chris, 96
assumptions
    communication and, 122
    complacency and, 97
    reasonable and professional assumptions, 96–97
ATP (adenosine triphosphate), 91
at-risk behavior, 95–96
Australian Geodetic Datum (AGD) 66 chart datum, 69
Australian Geodetic Datum (AGD) 84 chart datum, 69
Australia St. George Basin and Porosous Creek, 63–71, **66**, **67**, **68**, **70**, 89–90, 158
automatic radar plotting aids (ARPAs), 3, 40, 43, 71
autopilot, 71, 107

B
ballasting procedures, 151–**53**, 154, 156, 157, 158
Balsa 37
    accident and transition failure, 100–102, **101**
    bridge team roles and responsibilities, 132
    change of plan transition, 105
    changes in equipment settings, 107
    communication aboard, 111–13, 116, 120
    conferring and decision making, 138
    error chain aboard, 148, 157, 158–59
    hierarchy and yes-manship, 121–22
    pilot integration on, 104
Barge 101, **33**–**34**
beam bearing, **44**, 45, 48
berth-to-berth planning, 17–18, 41–42
bias, 120–22
Block Island, 105–6
boredom, 38–39
bottlenecks, 106
Bourbon Dolphin, 103–4
bow doors
    closing of, ix–xi, 5, 145, 149
    indicator lights, x, 149–50
    SOPs for closing, 7
bridge, layout and location of, 2–3

bridge resource management/bridge team management (BRM)
  accidents and incidents and, 4
  applicability of principles, 2–4
  attitude toward, 1
  contradictions, appearance of, 160
  goal of, 1–2
  principles of, xi
  rational behind, 109
  real-world situations and, 4
  relationships between elements of, 2
  training and certification in, 1, 12
  training in, purpose of, 1, 149, 150
  understanding and implementing, 159–60
bridge team
  Canadian regulations for, 74–75
  distraction and, 78
  getting underway transition, 106
  pilotage requirements, 11
  roles and responsibilities of, 8–11, 131–33
  structure of, 3, 72
  training of, 149
  watch conditions and, 13–15
Bristol Channel, 34–38, **35**, **36**
British Columbia Inside Passage
  Queen of the North incident, 77, 78–80
  Safari Spirit incident, 22–25, **23**, 31, 50, 58
buoyage systems, 115

C

caffeine, 92
Cape Henlopen, 89, 92
Capt Fred Bouchard
  accident and transition failure, 100–102, **101**
  change of plan transition, 105
  communication aboard, 111–13, 120
  error chain aboard, 148, 157
Caribbean Sea, 140
CEMS (Crew Endurance Management System), 93
Certificate of Inspection (COI), 12
chain spotting, 146–47, 148
Champlain, Samuel, 140
channel changes, 106
charts
  annotations to, 42–53, **44**
  chart datum, 42, 57, 69
  ECDIS, 3, 64, **124**, **126**
  electronic charts, 42, 47, 51; dimming screen of, 72, 75
  electronic chart systems (ECS), overreliance on, **64**–71
  inventory of, 42
  lack of, 81
  margins of safety, **41**, **44**, 45, 46–47
  modification of, 25
  no-go zones, **41**, 43, **44**, 46
  nonverbal communication on, 115
  paper charts, 42, 51
  raster charts, 70–71
  scale changes, notations about, 51
  scale of and passage planning, 22, **23**, 24, 42
  tailoring chart to needs of vessel, 42–53, **44**
  track lines, **41**, **44**, 45, 46, 48, 49–50
  transitions between charts, 51, 106–7
  vector charts, 72, 106
  waypoints, **41**, **44**, 45, 49–50
checklists, 15–16, 103
Chile, 42
chronic stress, 84–85
circadian rhythms, 90–**91**
clearing bearing, **44**, 45, 48
Clements Reef, **33**–34
closest points of approach (CPAs), 13, 71, 78, 133, 156, 159
Coast Guard, U.S.
  blood alcohol level standards, 89
  Cowslip and fog, 106
  Crew Endurance Management System, 93
  Cuyahoga incident, 135–36
  Empress of the North grounding, 127
  fatigue-related accidents, study of, 89
  Morro Bay and fog, 105–6
  North Cape incident and, 20
  river conditions and safety restrictions, 26, 28
  Valour rescue, 155–56
Coast Pilots, 56
COI (Certificate of Inspection), 12
collisions
  change of plan transition and, 100–102, **101**, 105
  COLREGS and avoidance of, 11–12
  VHF and text assisted collision, 116
COLREGS (International Regulations for Preventing Collisions at Sea), 11–12, 97, 115, 116, 127, 157
Columbia River, 106
command presence, 144
commission, errors of, 146
communication
  ambiguity in, 157
  assumptions and, 122
  breakdown in, xi, 120–22
  complacency and, 121
  components of communication system, 113
  contact information on passage brief, 58
  crew and pilots and, 104
  as distraction, 116
  distraction and, 121
  docking procedures and, 111
  effective communication, challenges of, 111
  effective techniques, 116–20, **117**, 122, 160
  external communication, 113
  failure of, 111–13, 122, 156
  fatigue and, 90, 121
  hierarchy and, 120, 121–22
  importance of, 117
  internal communication, 113
  language and cultural barriers, 97, 104, 112–13, 115–16, 118
  leadership and, 2
  lines of communication, 113–16

local vernacular, 115–16, 118
operating procedures and, 2
planning conference, 30, 32
proactive, x, 2, 116
protocols for, 111, 118–19
situational awareness and, 120–21
standing orders, 13, 125, 126
stress and, 121
style and structure, 119–20
teams and teamwork and, 2, 133–36
technology and, 111
terminology and, 118
working relationships and, 122
compass course, discrepancies in and confirmation of, 50
compasses
   drift or failure of, 50
   lack of, 81
complacency
   assumptions and, 97
   at-risk behavior and, 95–96
   characteristics associated with, 95
   communication and, 121
   distraction and, 79, 99
   effects of, 1, 5
   error chains and, 147–48, 156
   examples of, xi, 97–98
   experience and, 160
   fatigue and, 99
   ladder of inference and, **96**, 98
   leadership and, 99
   locked mental model and, 98
   monitoring stage of passage planning and, 37–38
   overreliance and, 63
   perspective of outside stranger and, 99
   prevention and deterrence of, 98–99
   repetition and, 95, 99
   situational awareness and, 62, 95, 100, 158
   stress and, 81, 99
   susceptibility to, 95
   transition and, 102
   understanding of, 95
conclusions, drawing, **96**
conferring
   decision-making and, 30–32, 138
   passage planning stage, 18, 28, 30–32
   shipboard culture for, 99
   transition and, 100–102
coning of attention, 38, 77
contingency planning, 17, 42, 53, 73, 75, 127
courses
   channel changes, 106
   course change, failure to make, 72–76, 89, 103
   course change, three-way meetings and, 100–102, 157
   maneuvering and subtle course changes, 157
   transposing course digits, 50

courtesy, 120
Cowslip, 106
CPAs (closest points of approach), 13, 71, 78, 133, 156, 159
crew
   continuity and experience of, 85
   experience of and margins of safety, 47
   manning adjustments, chart notations about, 52
   mental model, shared, 6, 102, 132, 136
   personality clashes, 86
   relationships between, 74, 75–76, 104
   situational awareness and turnover of, xi, 85
   strengths and weaknesses of, 3
   structure of bridge team, 3
   training of, 85–86
   turnover of, x–xi, 85, 103–4, 149
   working relationships and communication, 122
   workload of, stress and, 86
Crew Endurance Management System (CEMS), 93
crossing situations, 116
cross-referencing, 39–40, 57–58, **65**, 66–71
cross-track error, 45, 48, 49–50
cultural barriers and bias, 97, 104, 112–13, 115–16, 118, 122
culture of safety, 1, 12
Cuyahoga, 135–36

D

danger bearings, 48
day into night transition, 105
decision making
   ambiguity and, 157
   assumptions and, 96–97
   bad decisions
      anatomy of, 139–41
      defense against, 30
      recognition of, 137
   conferring and, 30–32, 138
   default decisions, 139–40
   effective decision making, deterrents to, 137–39
   fatigue and, 139
   judgment, defense against poor, 30
   ladder of inference and, **96**
   leadership and, 137, 143
   momentum, judgment, and, 159
   process for, 137
   timing of decisions, 137
   training and, 139–40
   uncertainty and, 140–41
default decisions, 139–40
delaying, 53
delegation, 143
depths
   depth verification points, **44**, 45, 49
   forward-scanning sonar, 67, 158
designated leadership, 141–42
distance off, 48
distance of passage leg, recording of, 50

distraction
    accidents and incidents and, xi
    awareness of, 72
    communication and, 121
    communication as, 116
    complacency and, 79, 99
    crew relationships and, 74, 75–76
    defenses against, 78–80
    effects of, 1
    failure to pay attention, 73–74
    fatigue and, 2, 79
    fixation, 77
    personal distractions, 77
    professional distractions, 77–78
    Queen of the North example, **72**–76, 77, 78–80
    situational awareness and, 62, 100, 158
    stress and, 2, 79, 81
    structural distractions, 78
    switchtasking, 76–77
    timing of, 78, 80
    transition and, 76, 79
docking procedures, communication and, 111
draft, logging of, x, 11
drug and substance abuse, 75, 86, 143

E
Eads Bridge, 27–30, **28**, **29**, 53
EBLs (electronic bearing lines), 25
Egmont Channel and Tampa Bay, 100–102, **101**, 105, 106, 111–13, 116, 121–22, 132, 138, 148, 157, 158–59
Elaine G, 103
electronic bearing lines (EBLs), 25, 49
electronic chart display and information system (ECDIS), 3, 64, **124**, **126**
electronic chart systems (ECS), overreliance on, **64**–71
emotional intelligence, 143
Empress of the North
    accident and teamwork failure, 123–28, **124**, **126**, 129
    bridge team communication, 134–35, 136
    bridge team roles and responsibilities, 131–32, 133
    decision-making on, 137, 138, 139, 140–41
    error chain aboard, 157–58, 159
    leadership on, 142, 143
engines
    failure of, 37, 95
    fire in engine room, 19–**20**, 21, 95, 150
entertainment and music, 76, 77
environmental conditions. *See also* fog; night operations; visibility; weather conditions
    communication and, 116
    fatigue and, 88
    passage planning and, 18, 32, 56
    situational awareness and, 159
    watch conditions, 13–15
environmentally sensitive areas, 46

equipment and instruments
    changes in equipment, 107
    changes in settings, 107
    communication devices, 114
    cross-referencing, 39–40, 57–58, **65**, 66–71
    examples of, 3
    familiarity with, 126
    overreliance on, 63, 71
    readiness and status of, 51–52
    reliability of, 71
    single-point errors, 6
    transitions and situational awareness, 106–7
    use and reliance on, 3
errors. *See also* accidents and incidents
    accident cascades, 150, 154–56
    consequences of, 145
    cross-track error, 45, 48, 49–50
    error chain, failure to break, 156–59
    error chains, 146–**48**, 150, 160
    error trapping, 149–50
    ghost errors, 145
    GPS errors and blunders, 70
    human errors, predictability of, 160
    human errors, types of, 145–50, **147**, **148**
    prevalence of, 145
    single-point errors, 6, 24, 134–36
    Valour error chain example, 150–56, **151**, **152**, **153**, **155**, 157, 158, 159
ETAs, 26, 57
ethics and ethical behavior, x, 143–44, 156
evacuation plans, 73, 75, 127
Ever Grade, 106
execution stage of passage planning, 18, 26–30
experience
    complacency and, 160
    of crew, 47, 85
    decision-making and, 137, 138
    situational awareness and, 148
    for watchkeeping, 123–24, 126–27, 128, 137
Exxon Valdez, 50, 149

F
face-to-face communication, 113–14, 119–20
Fantome, 140
fatigue
    accidents and incidents and, xi, 88–90
    blood alcohol comparisons, 89, 90
    circadian rhythms, 90–**91**
    cognitive impairment from, 89–90
    communication and, 121
    complacency and, 99
    concern about, 88
    decision making and, 139
    definition of, 89–90
    distraction and, 2, 79
    effects of, 1
    falling asleep during watchkeeping, 88
    fitness for duty standards, 92–94

management of, 90–92, **91**, 94
measurement of, 88
prevalence of, 94
red zone, **91**
regulations related to, 92–94
self-reporting of, 89, 90
signs of, 90
situational awareness and, 61, 62, 100, 158
sleep and, 91–92
stress and, 86, 92
Swiss cheese error model and, 149
tiredness and, 89–90
understanding of, 88
work-rest balance, 92
Faulkner, William, 121
fire aboard *Scandia*, 19–**20**, 21, 95, 150
fitness for duty standards, 92–94
fixation, 77
fog
　error chain and, 159
　good to restricted visibility transition, 105–6
　Long Island Sound ferry collision, 89
　meeting situations and, 106
　Mobile River incident, 81–83, **82**, 98, 105
　Ohio River collision, 103
　speed during fog conditions, 159
forehandedness, 78
foreshadowing, 147–48
forward-scanning sonar, 67, 158
functional leadership, 141–42

G

general adaptation syndrome (GAS) stress model, 83–**84**
generational bias, 122
ghost errors, 145
goals, incompatible, 157–58
GPS
　antennas, 69
　errors and blunders, 70
　monitoring and cross-referencing, 57–58, **65**, 66–71
　position fixing with, 57
　WGS 84 datum and, 42
groupthink, 6

H

habitual at-risk behavior, 95–96
head marks, **44**, 45, 48
Herald of Free Enterprise
　at-risk behavior and bad luck, 95
　bridge team roles and responsibilities, 132
　communication on, 120
　complacency and, xi, 97
　crew turnover on, x–xi, 85, 103
　distraction on, 78
　error trapping on, 149–50
　getting underway transition, 106
　ISM Code in response to sinking of, 12

　latent conditions aboard, 145
　leadership on, 141
　schedule changes and management goals, ix–x, 104–5, 157
　sinking of, procedures that lead to, **viii**–xi, 145, 150
　SOPs on, 5, 7–8, 11, 16
　unethical behavior on, 144
　written communication on, 115
hierarchy, 120, 121–22
human factors
　accidents and incidents and, xi, 1–2, 4
　awareness of problems related to, 1
　situational awareness and, 61–62, 109
human interaction, mistakes and, 109
hurricanes, 140
hydration, fatigue and, 92
Hyuandi Dominion, 116

I

imagination, decision-making and lack of, 138–39
IMO (International Maritime Organization), 18, 92–94
inference, ladder of, **96**, 98
information, decision making and lack of, 137–38, 156
intentional at-risk behavior, 95
intentional teams, 123
internal SOPs, 12–15
International Convention for the Safety of Life at Sea (SOLAS), 17, 64
International Management Code for the Safe Operation of Ships and for Pollution Prevention, 12
International Maritime Organization (IMO), 18, 92–94
International Regulations for Preventing Collisions at Sea (COLREGS), 11–12, 97, 115, 116, 127, 157
International Safety Management (ISM) Code, 12
interpersonal skills, 142

J

Jambo, 89, 90, 91
judgment
　defense against poor, 30
　momentum and, 159
Justine Foss, 155–56

K

Kisameet Bay, **22**–25
Klein, Gary, 137, 138

L

ladder of inference, **96**, 98
lake levels, 56
Lake Michigan, **44**–45
La Moure County, 42
language and cultural barriers, 97, 104, 112–13, 115–16, 118
latent conditions, 145–46, 150, 156
leadership
　ability for, 141
　assignment of responsibilities, 141
　changes in, 103–4

characteristics of effective leaders, 141, 142, 144
communication and, 2
complacency and, 99
decision making and, 137, 143
designated leadership, 141–42
dysfunctional working relationships, x, xi
failure of, x, xi
functional leadership, 141–42
importance of, 141
persona, 144
shared leadership, 133
situational awareness and, 142
sources of, 141
of teams, 2, 133
techniques, 142–44
types of leaders, 141, 156, 157
lee shores, 36–37, 46, 95
life jackets and survival suits, 154, 156
life rafts, 20, 75, 102, 127, 155–56
limited tonnage vessels, 2
list
ballast tanks and, 151
liquid ballast to correct, 151–53, 154, 156, 157, 158
permanent list, 151
stability letter and, 151
locked mental model, 98, 136, 138, 160
Long Island Sound, 89, 92
lookouts, 52, 100, 103, 127, 145–46
loop communication, **117**–18
luck
bad luck, 95–96, 146
better to be good than lucky, 2
good luck, 63, 145

M

machinery, readiness and status of, 51–52
maneuverability, margins of safety and, 46
margins of safety (MOS), **41**, **44**, 45, 46–47
Maria Asumpta
abort point, 53
accident and passage planning, **34**–38, 40
at-risk behavior and bad luck, 95
change of plan transition, 105
complacency and, 98
error chain aboard, 146, 157, 159
fixation of captain, 77
margins of safety, 46
stress of captain, 84
marine environment
accidents and incidents and, 149
risk management in, 4, 159
situational awareness and, 62, 149
master-pilot exchange (MPX), 104, 129–31, 161
Matia Island, **33**, 34

Mauvilla
accident and stress, 81–83, **82**, 84, 85
change of plan transition, 105
decision making on, 138
error chain aboard, 159
error trapping after incident with, 149
mental model of pilot, 98
speed aboard, 159
workload of crew, stress and, 86
meals, fatigue and, 92
medications
alertness and sleep quality and, 92
self-medication and substance abuse, 75, 86, 143
meeting situations
course change and three-way meetings, 100–102, **101**
fog and, 106
port-to-port meeting, 85, 97, 100–101, 112–13, 116, 118, 136
starboard-to-starboard meeting, 95, 101, 105, 116, 138
mental model
locked mental model, 98, 136, 138, 160
shared mental model, 6, 102, 132, 136, 160
standard operating procedures and, 5–6, 98
transition and, 102
Mercury
accident and passage planning, **33**–34, 37–38, 42, 49, 58, 145
complacency and, 98
margins of safety, 46
SOPs on, 156
micromanagement, 143
mindfulness, 78
mission teams, 129–31
Mississippi River, 26–30, **27**, **28**, **29**
Mobile River, 81–83, **82**, 138
momentum, judgment and, 159
Mona Island, 88–89
monitoring stage of passage planning, 18, 32–40
Morro Bay, 105–6
motivation, 142
Mouls, **36**–37
MPX (master-pilot exchange), 104, 129–31, 161
multitasking, 76, 86
music and entertainment, 76, 77

N

napping, strategic, 91, 92
natural ranges, **44**, 45, 47, 48
navaids (aids to navigation), 43
navigation
margins of safety and, 47
navigational control, maintenance of, 2
position information sources, 2
on rivers, 26
by sight and radar, 81
single-point errors, 24
standing orders, 13, 125, 126
technology and automation of, 71
transition and navigation changes, 106
Navigation Rules, 11–12, 97, 115, 116, 127, 157

night operations
    day into night transition, 105
    electronic charts, dimming screen of, 72, 75
    error chain and, 159
    margins of safety and, 46
    visibility and, 28, **41**
night orders, 13
no-go zones (NGZs), **41**, 43, **44**, 45, 46
nonverbal communication, 115–16
North Cape, 18–21, **20**
North Star, 89

O

obstructive sleep apnea (OSA), 92
Ohio River, 103
omission, errors of, 146
OSA (obstructive sleep apnea), 92
overreliance
    complacency and, 63
    definition and characteristics of, 63
    on equipment and instruments, 63, 71
    on resources, 63, 71, 160
    situational awareness and, 62, 100, 121
    on technology, 63
    True North example, 63–71, **64**, **65**, **66**, **67**, **68**, **70**, 90, 95
oversight, 143
overtaking maneuvers, 100–102, 106, 111–13

P

paperwork, 78
parallel indexing, **44**, 45, 48–49
passage planning. *See also* charts
    ambiguity in, 126
    appraisal stage, 18–21
    benefits of, 1, 2, 17, 58–59
    berth-to-berth planning, 17–18, 41–42
    change of plan transition, 105
    complacency, deterrence of, 99
    conferring stage, 18, 28, 30–32
    contingency planning, 17, 42, 53
    cross-referencing, 39–40, 57–58
    departure from, 156–57
    execution stage, 18, 26–30
    human errors and, 156
    improvement to approach to, 149
    machinery and equipment readiness, 51–52
    monitoring stage, 18, 32–40
    optimum route, crafting the, 47–53
    passage brief, **41**, 53–58, **54**, **55**
    planning conference, 30, 32
    planning stage, 18, 21–25
    purpose and requirements of, 17–18
    risk management through, 17
    safe water and route, **41**, **44**, 47
    situational awareness and, 17
    STCW requirement, 18, 42
    time and place, stress and, 85
    tools and tactics, **41**–42, 58–59

personal distractions, 77
pilots
    integration of, 104
    master-pilot exchange (MPX), 104, 129–31, 161
    ship casualties and, 104
    on towboats, 26, 81
planning and procedures. *See also* passage planning; standard operating procedures
    complacency, deterrence of, 99
    contingency planning, 17, 42, 53, 73, 75, 127
    distraction and, 78
    failures of, 5
    following, 5
    mental model and, 5–6
    value of, 5
planning conference, 30, 32
planning stage of passage planning, 18, 21–25
port-to-port meeting, 85, 97, 100–101, 112–13, 116, 118, 136
position
    communication of, 119
    relative position to objects, system to identify, 119
position fixing
    frequency of, **41**, 58, 156
    methods of, **41**, 57–58
    monitoring and cross-referencing tasks, 39–40, 57–58
    position-monitoring tools, **44**–45, 47–50
Potomac River, 135–36
prévoyance, 140, 147
priorities
    distraction and, 78, **79**, 80
    incompatible goals, 157–58
    resources and, 4
    responsibilities and duties and, ix–x, 86
    time-use matrix, 79, 86
    transition and, 100
professional bias, 121
professional distractions, 77–78
professional imagination, 138–39
Puget Sound, **33**–34, 42, 145

Q

Queen of the North
    accident and distraction, **72**–76, 77, 78–80
    changes in equipment, 107
    error chain aboard, 146–47
    transition on, 103
    unethical behavior on, 144

R

radar
    ARPAs, 3, 40, 43, 71
    electronic bearing lines (EBLs), 25, 49
    monitoring and cross-referencing tasks, 40
    navigation by, 81
    parallel indexing, **44**, 45, 48–49
    position fixing with, 57

training for use of, 83, 149
variable range marker (VRM), 23, **24**, 25, 49, 95
railroad bridge incident, 81–83, **82**
raster charts, 70–71
Reason, J. T., 145, 148
redundancy, 6
red zone, **91**
regulatory obligations and warnings, 52
remote areas, operating in, 22, 24, 25
repetition, complacency and, 95, 99
resources
   definition and examples of, 3–4
   experience and mistakes as, 3–4
   lack of or loss of, 158
   management of, 4, 160
   overreliance on, 63, 71, 160
   priorities and, 4
   single-point errors, 6
   stress and inadequate resources, 85
respect, 120, 142
responsibilities and duties
   attitude toward, ix, xi
   of bridge team, 8–11, 131–33
   priorities and, ix–x, 86
   transfer of, ix
   watch conditions and, 13–15
   zones of responsibility, x
restricted areas, 46
Rhode Island
   fishing industry in, 20
   oil spill on coast of, 18–21, **19**
risk management
   at-risk behavior, 95–96
   in marine environment, 4, 159
   perspective of outside stranger, 99
   situational awareness and, 61
   through passage planning, 17
rivers
   currents in, 26–30, **31**, **41**
   error chain and conditions in, 159
   flood conditions, 27–30, **28**, **29**
   navigation on, 26
   operations on, 26, 31
   pop rise conditions, 29
   stages and passage brief, 56
Rocky Island grounding, 123–28, **124**, **126**, 157
Royal Majesty, 58
Rules of the Road, 11–12, 97, 115, 116, 127, 157
run times, 33–34, 57

S

Safari Spirit
   accident and passage planning, 22–25, **23**
   at-risk behavior and bad luck, 95
   complacency and, 97
   conferring and decision-making, 31–32
   errors aboard, 147–48

   error trapping after incident with, 149
   getting underway transition, 106
   position fixing procedures, 50, 58
safety management system (SMS), 12
safe water and route, **41**, **44**, 47
St. Louis, 26–30, **28**, **29**
Scandia
   abort point, 53
   accident and passage planning, 18–21, **19**, **20**, 56
   at-risk behavior and bad luck, 95
   complacency and, 97
   error chain aboard, 158, 159
   error trapping on, 150
   fire aboard, 150
   margins of safety, 46
schedule changes and delays, ix–x, 53, 104–5, 139, 157
Scotland, 89
Seafarer
   accident and transition failure, 100–102, **101**, 138
   change of plan transition, 105
   communication aboard, 111–13, 116, 120
   error chain aboard, 148, 157
   hierarchy and yes-manship, 121
   transition on, 103
self-awareness, 87, 142, 144
self-medication and substance abuse, 75, 86, 143
Selye, Hans, 83–84
shared mental model, 6, 102, 132, 136, 160
ships, boats, and vessels
   BRM principles, applicability of, 2
   purpose-built vessels, 18
   small ships, definition of, 2
shoreside management
   accidents and incidents, responses to, 149
   error trapping, recognition of benefits of, 150
   institutional changes, ix–x, 104–5
   leadership from, 141
   priorities and incompatible goals, 157–58
   problems related to, xi
   realistic procedures and operations, x
signal flags and devices, 116
single-point errors, 6, 24, 134–36
situational awareness
   accidents and, xi
   chain spotting and, 148
   change of watch and, 103
   characteristics of, 61
   checklists and, 15
   communication and, 120–21
   complacency and, 62, 95, 100, 158
   crew turnover and, xi, 85
   definition of, 3
   degrees of, 61
   distraction and, 62, 100, 158
   environmental conditions and, 159
   experience and, 148
   factors that undermine, 61–62, 109, 148

fatigue and, 61, 62, 100, 158
foreshadowing and, 148
human factors and, 61–62, 109
information sources and, 3
leadership and, 142
marine environment and, 62, 149
navigation transitions and, 106
overreliance and, 62, 100, 121
passage planning and, 17
pilots and, 104
risk management and, 61
speed and, 159
stress and, 62, 81, 83, 87, 100, 158
teams and teamwork and, 123
technology and, 61–62
transition and, 62, 100, 106–7, 121
Sky Hope, 116
sleep cycles and sleep fragmentation, 91
sleep debt, 92
sleep deprivation, 91
sleep disorders, 92
SMS (safety management system), 12
social bias, 122
SOLAS (International Convention for the Safety of Life at Sea), 17, 64
sonar, forward-scanning, 67, 158
sound signals, 116
Sources of Power (Klein), 137
speed
  margins of safety and, 46
  passage brief and speed plan, 56–57
  safe speed, 56–57
  situational awareness and, 159
  speed change chart annotations, **44**, 45, 51
  three-way meetings and, 100–102
  time and speed, stress and, 85
spoken communication, 113–14
spontaneous teams, 123
stability, trim and, x
stability letter, 12, 151, 152, 156, 157
standard operating procedures (SOPs)
  adherance to, 11
  adherence to, 8, 12, 16
  Andrew J. Barberi, SOPs on, 8–11, 16
  benefits of, 1, 7, 16
  breakdown of, xi
  checklists, 15–16
  clarity of, 8, 11
  communication and, 2
  complacency, deterrence of, 99
  criteria for effective procedures, 7–8
  departure from, 156–57
  enforcement of, x
  establishment of, 11–16
  evolution of, 7
  flawed procedures, x
  Herald of Free Enterprise, SOPs on, 5, 7–8, 11, 16

human errors and, 156
internal SOPs, 12–15
mental model and, 5–6, 98
priorities and, ix–x
professionally sound procedures, 7, 16
realistic procedures, x
regulatory sources of, 11–12
standardization of, ix, 2, 7
training on, 11
Standards of Training, Certification, and Watchkeeping (STCW) convention
  compliance with, 1
  fitness for duty standards, 92–94
  lookout requirement, 127
  Manila Amendments to, 46
  passage planning requirement, 17, 42
  purpose and requirements of, 12
standing orders, 12–13, 125, 126
standing teams, 128–29
starboard-to-starboard meeting, 95, 101, 105, 116, 138
steering light, 26, 29, 30
steering system, operation of, 72, 73, 75
stern marks, **44**, 45, 48
strategic napping, 91, 92
stress
  accidents and incidents and, xi
  acute stress, 84, 92
  chronic stress, 84–85
  communication and, 121
  complacency and, 81, 99
  distraction and, 2, 79, 81
  effects of, 1
  fatigue and, 86, 92
  general adaptation syndrome stress model, 83–**84**
  management of, 85–87
  as natural response, 83
  normal stress, 81
  overloading, 81
  personal stress, 86
  professional stress, 86
  self-awareness and, 87
  signs of, 84
  situational awareness and, 62, 81, 83, 87, 100, 158
  stressors, 83
  technology and, 86
  training and, 85–86
  work-play balance and, 86–87
structural distractions, 78
substance abuse, 75, 86, 143
Sunset Limited accident, 82–83, 149. See also *Mauvilla*
survival suits and life jackets, 154, 156
Swiss cheese error model, 147, **148**–49, 157
switchtasking, 76–77

T

Tampa Bay and Egmont Channel, 100–102, **101**, 105, 106, 111–13, 116, 121–22, 132, 138, 148, 157, 158–59

tanks, cross-connections between, 151, 152, **153**, 154
team-of-one, 3, 38–**39**, 76, 129, 134–35
teams and teamwork. *See also* bridge team
    attributes of a team, 131–36, 160
    benefits and importance of, 123, 134, 136
    communication and, 2, 133–36
    complacency and, xi
    composition of teams, 131–32
    culture of violation, x, 157
    dysfunctional working relationships, x, xi
    failure of, 123, 127–28
    intentional teams, 123
    leadership of teams, 2, 133
    management of teams, 4
    mental model, shared, 6, 102, 132, 136
    quality of teamwork, 136
    roles and responsibilities in, 123, 126–27, 128, 131–33, 136
    self-sufficiency and, 129
    situational awareness and, 123
    spontaneous teams, 123
    watchkeeping teams, types of, 128–31
technology
    communication and, 111
    navigation, automation of, 71
    overreliance on, 63
    reliance on, 1
    situational awareness and, 61–62
    stress and, 86
terminology
    communication and, 118
    standardization of, 7
Thomas B. McCabe, 81, 82
tides and currents
    error chain and, 159
    passage brief and, 56
    passage planning and, 22, 23, 25, **41**
    in Puget Sound, 33–34
tight quarters, 22–**23**, 25
time management, 79, 86
tiredness and fatigue, 89–90
towboats. See also *Anne Holly*; *Mauvilla*; *Scandia*
    assist vessels for, 28, 30, 158
    length of tows, 26, **27**
    margins of safety for, 46
    operations on, 26
    steering light, 29, 30
    training of operators, 83
track lines, **44**, 45, 46, 48, 49–50
traffic monitoring
    margins of safety and, 46–47
    standing orders, 13
    time and place, stress and, 85
    transition and, 100–102, **101**
    vessel traffic service, 3, 33, 58, 72
transition
    accidents and incidents and, 100–102, **101**, 111–13, 158–59
    between charts, 51, 106–7
    complacency and, 102

    conferring and, 100–102
    definition of, 100
    distraction and, 76, 79
    equipment and instruments and situational awareness, 106–7
    management of, 102–7
    mental model and, 102
    priorities and, 100
    situational awareness and, 62, 100, 106–7, 121
    switchtasking and, 76
trim, stability and, x
True North
    at-risk behavior and bad luck, 95
    complacency and, 98
    decision making on, 138
    error chain aboard, 158
    fatigue and, 89–90
    overreliance and, 63–71, **64**, **65**, **66**, **67**, **68**, **70**, 90
turning range, **44**, 45, 49
Twain, Mark, 98
twilight, 105

U

uncertainty, 140–41
under-keel clearances (UKCs), **41**, 43, 44, 46, 50
unintentional at-risk behavior, 95

V

Valour, 150–56, **151**, **152**, **153**, **155**, 157, 158, 159
variable range marker (VRM), 23, **24**, 25, 49, 95
verbal communication, 113–14, 116
vessel traffic service (VTS), 3, 33, 58, 72
VHF radio
    changes in settings, 107
    collisions and communication over, 116
    contact information on passage brief, 58
    monitoring of, 52
    proactive use of, 116
visibility
    error chain and, 159
    good to restricted visibility transition, 105–6
    lookouts, 52, 100, 103, 127, 145–46
    night operations and, 28, **41**
    signal flags and devices and, 116
    stress and, 81–83, **82**

W

watch conditions, 13–15
watchkeeping
    assumptions and, 97
    bridge and, 2–3
    change of watch, 100, 102–3
    cross-referencing, 39–40, 57–58, **65**, 66–71
    environmental conditions and error chains, 159
    experience for, 123–24, 126–27, 128, 137
    failure of and accidents, xi
    falling asleep during, 88
    logkeeping, 34, 39
    music and entertainment during, 76, 77

procedures that enhance, 5
single-point errors and, 6, 134–36
SOPs for, 11–16
standing orders, 125
team-of-one, 3, 38–**39**, 76, 129, 134–35
two-watch system, 3, 26, 28, 31
types of teams for, 128–31
watch conditions, chart notations about, 52
watch rotation, 89, 91, 139
watertight subdivisions, 75
waypoints, **44**, 45, 49–50
weather conditions
    appraisal stage of passage planning and, 18–21
    error chain and, 159
    margins of safety and, 46
    passage brief and, 56
    passage planning and, **41**
    reliability of weather forecasts, 21
    vulnerability of decisions and, 21
weather gauge, 85
WGS (World Geodetic System) 84 chart datum, 42, 69, 71
wheel-over points, **41**, 50
Winters, Richard, 144
workload of crew, stress and, 86
World Geodetic System (WGS) 84 chart datum, 42, 69, 71
written communication, 114–15

Y

yes-manship, 121–22

Z

Zeebrugge, Belgium, ix, x, 104–5